Vasa

Vasa

FRED HOCKER

Photographs by
ANNELI KARLSSON

MEDSTRÖMS BOKFÖRLAG

OXBOW BOOKS

Copyright © 2011 Medströms Bokförlag and National Maritime Museums in Sweden

All rights reserved. No part of this publication may be reproduced, stored in a retrieval system, or transmitted in any form or by any means, electronic, mechanical, photocopying, recording or otherwise, without the prior permission of the copyright owner.

Cover: photos by Anneli Karlsson, National Maritime Museums in Sweden
Graphic design: Annika Lyth and Mark Evans
Cover design: Otto Degerman

Printed in Sweden by Elanders Fälth & Hässler, Mölnlycke, 2011

Distributed outside Sweden and Finland by Oxbow Books

ISBN 978-91-7329-101-9

This book is available from:

Medströms Bokförlag
Thomas Magnusson
Artillerigatan 13
Östra Blockhuset
114 51 Stockholm
Sweden
Tel: +46 (0)8 411 33 72
E-mail: thomas.magnusson@medstromsbokforlag.se

Oxbow Books
10 Hythe Bridge Street
Oxford, OX1 2EW
United Kingdom
Tel: +44 (0)1865 241249
E-mail: oxbow@oxbowbooks.com
www.oxbowbooks.com

The David Brown Book Co
PO Box 511
Oakville CT 06779
USA
Toll-free: 800 791 9354
Tel: 860 945 9329
E-mail: queries@dbbconline.com

Contents

- 7 FOREWORD
- 9 A SHIP'S TALE
- 19 AN AGE OF FAITH AND VIOLENCE
- 33 TIMBER, IRON AND MEN
- 49 THE MACHINE OF WAR
- 67 THE SYMBOLIC WARSHIP
- 83 THE FULL-RIGGED SHIP
- 99 A FLOATING COMMUNITY
- 121 SINKING
- 143 CONSEQUENCES
- 157 LOST BUT NOT FORGOTTEN
- 171 SALVAGE AND EXCAVATION
- 195 VASA TODAY
- 204 SOURCE MATERIAL
- 208 PICTURE SOURCES
- 208 INDEX

Foreword

The *Vasa* story has many dimensions. There is the grand fiasco, the hugely expensive ship which sank so annoyingly just a few short minutes into her maiden voyage. But there is also *Vasa* as a key to understanding the 17th century as a constant battle for resources and control of trade routes, of the navy as a central element and of the rapid construction of a strong Swedish nation-state. The people on board *Vasa* tell us about life in the 17th century: food shortages, disease, strong faith in God and a tight community with violent traits. In terms of military history, *Vasa* represents an interesting hybrid between two different types of warship, an older version built for transporting troops, and a younger which actively participated in sea battles – two purposes difficult to combine, and one of several explanations why *Vasa* was built with less stability. The 20th-century *Vasa* story is no less exciting – the salvage took place in a country dominated by a belief in the power of technology, with Anders Franzén in the centre of male networks and informal decision-making.

Vasa tells us about the history of Sweden, both long ago and quite recently. But we have to know what questions to ask, to get at that history – something not easily done.

When Fred Hocker started work on understanding *Vasa* and her many stories in 2003, there had already been substantial research done by other scholars. Now he brought a new dimension to the *Vasa* research. By combining his skills as an archaeologist and historian, and by knowing so much about how 17th-century vessels were constructed, Fred Hocker has dug deeper into the history of the ship and given us a completely new story of *Vasa*, a dramatised tale based on extensive scientific research. And thus, this book also gives us a new story of Sweden.

Stockholm, August 2011

Marika Hedin, Director, Vasa Museum

Military power and a kingdom's expensive pride: one of the enduring monuments of the Swedish Great Power period is also a symbol of folly and disaster. One of the ship's gunports shows a snarling lion, fearsome but now impotent.

Vasa now lies in permanent winter ordinary. The Swedish fleet sailed only part of the year, due to ice and storms, and spent the cold months at the Stockholm navy yard with most of the rig taken down and stored, as the ship is displayed today.

A ship's tale

1628

It felt wrong. He had been in the navy long enough to know that. All ships heeled when the wind filled the sails, but this was wrong. There was no life in the movement. The ship should be straining to right herself, but rolled drunkenly. She hung there for a long moment before seeming to decide what to do. She came back up slowly, too slowly, as the current carried her farther down the harbour. She was not answering the helm very well, drifting more than sailing. She was caught in a dangerous paradox: there was not enough wind to move her much faster than the current, so there was no real water flow over the rudder and she was hard to steer in the confined waters of Strömmen, but if the wind increased enough to make her sail, she would heel over until the gunports were in the water. Would she manage to run aground before she capsized?

A month ago, he had been one of thirty men commanded to run back and forth across the upper deck to make the ship roll. The yard captain supervising the ship's construction knew that the ship was not safe and had organized a demonstration of her lack of stability for the vice admiral, but the admiral had stopped the demonstration, afraid the ship would roll over at the quay. Since that day there had been a heavy blanket of foreboding lying over the preparations for sailing. Combined with the pressure to get the ship to sea, it had everyone on edge.

He was still below decks, detailed to clear up after the salute they had fired, stow the rammers and ladles and coil down the gun tackles. He could see through the open gunports that they were still close in to the cliffs, but they were coming up to the gap at Tegelviken. If there was any wind out here, it would funnel down through that gap in a rush. There were ruffles on the water. Without thinking about it, he caught hold of a ringbolt and braced himself against the nearest cannon.

When the gust came, it filled the topsails with a snap. The ship heeled immediately, with that same lifeless roll as before, but with more wind this time, she kept heeling. The deck sloped away from under his feet, making it harder to

Water coming on board above the waterline should drain back off the ship through scuppers, small holes in the sides, only 65 mm in diameter. There were only three or four on each side of each deck, too few and too easily blocked.

Vasa's lower gundeck is strongly built, of oak, to carry the heavy guns of the main armament. It was also crowned and caulked, so that water would run off to the sides.

1628

THE SHIP SINKS

Vasa heels 17 degrees to port and the catastrophe is certain. Water pours in through the open gunports on the lower gundeck and builds up on the deck. Soon it will rise to the hatches and begin to run down into the hold. Ladders are difficult to climb at this angle of heel, and loose objects slide and tumble to port. The cannon are secure, but pull at their breechings.

Vasa sank before the eyes of the people of Stockholm, and even after the ship was on the bottom, the tops of the masts were visible above the water.

stand. He heard an order shouted to haul in the guns and secure the gunports, but that would take far too long. He held tighter to the ringbolt and considered how to get out of the lower gundeck before it was too late. He could not go through the gunports, filled with cannon, so his best chance was the ladder by the mainmast up to the upper gundeck and from there to the upper deck and safety.

He saw the water pour over the sills of the midships gunports on the port side, and knew he had to move. As the water built up on the deck, it would push the port side farther down and the water would rush in even faster. The ship was doomed, and he hoped that he had not waited too long.

He took a deep breath and let go of the ringbolt, starting aft towards the ladder, but there was little grip on the tilting deck. The ship gave a lurch and he slid sideways, tripping over the corner of a hatch coaming. His own body weight was now his enemy, as he accelerated towards the port side with no way to stop himself. His scrabbling feet fetched up against the anchor cable laid out on the deck, in preparation for a mooring that would never happen. He bounced off the breech of a cannon, twisted, and pitched head-first into the oak ceiling next to a gunport. It was the last thing he knew. His feet slid under the gun and his legs jammed between the gun carriage and the anchor cable. Water poured in over him, and he probably never knew that he drowned, sinking into the darkness with the king's newest ship.

1958

The dive boss had told him, like all the other divers, that it was important to follow the underside of the ship when digging the tunnels. It would minimize the amount of clay that had to be removed, would help to break the suction between the old ship and the harbour bottom, and would reduce the risk of the tunnel caving in. It was dangerous work. He was at the bottom of a 5-metre shaft into the soft bottom, 32 metres below the surface. He could only work for an hour before a slow ascent and over an hour sitting in the decompression chamber, breathing out the nitrogen that had accumulated in his body.

He was well into the tunnel, trying not to think about the hundreds of tons of ancient, rotten wood over his head while he cut the glacial clay with a high-pressure water jet. The suction dredge jammed between his knees carried the mud away, back up the tunnel and to the surface. There was just enough room for his bulky diving equipment, the jet, the dredge and his air hose and telephone line, and he had to make sure to make the tunnel wide enough to allow him to move his arms. He had not been especially careful to clean the ceiling of the tunnel, formed by the bottom planks of the ship. It was impossible to see in the pitch dark with mud swirling around, and the diving helmet restricted the view above him in any case.

The water flowing into the tunnel to replace the water being sucked out by the dredge was eroding the clay he had left stuck to the bottom of the ship. At last it peeled away from the wood and fell into the tunnel. He felt it bury his legs. A cave-in, his worst fear – he was trapped in a small, dark hole at the bottom of the sea. At the surface, the dive boss could hear his suddenly fast, ragged breathing over the dive telephone, and he stepped quickly to the microphone.

The boss's gruff, matter-of-fact voice sounded in the copper helmet. "Diver! What is the matter?"

"The tunnel collapsed! I'm buried! I'm going to die!"

The dive boss was all business. "Are you breathing?"

"Yessir!"

"Then you're not going to die, we can pump air to you all day long. Are your dredge and water jet still working?"

"Yessir."

"Then turn around and dig yourself out. And next time, remember to keep the tunnel roof clear!"

"Uhh, yes sir." Feeling a little foolish, both for having caused the cave-in and for panicking, he set about working the jet around behind him and freeing his legs so that he could turn around and clear the rest of the tunnel. He should have listened to the dive boss; he was the most experienced salvage diver in Sweden and had been running this operation for more than two years without any accidents or injuries. Even the commodore in charge of the navy yard listened to the dive boss.

1961

It was not the first skeleton they had found since the excavation began two months earlier; counting the finds made by the divers during the salvage, it was the eighth. By now they had a good grip on the excavation process and things

The heavy divers of the 1950s had a dangerous and difficult job, wearing equipment which weighed nearly a hundred kilograms. Here, one of the salvage divers brings a sculpture from the stern of the ship to the surface.

"Gustafsson," still wearing his shoes, is excavated on the lower gundeck.

were running smoothly. Washing away the mud, removing artefacts, cataloguing them and packing them for transport to the conservation lab had settled into a routine. They had to work fast, to clear the ship of hundreds of tons of mud and objects as quickly as possible, since the engineers were worried about how long the ship could hold together under so much weight. Each day brought hundreds of objects to light, but there was no time to stop and wonder, or contemplate the meaning of what they were finding. But this was different.

The other human remains they had encountered were either single loose bones in the mud or jumbled piles of bones. This was still clearly a person, a man lying on his right side with his head against the side of the ship and his legs trapped under a gun carriage on the port side of the lower gundeck. His feet were still in his shoes, and he still wore the remains of a jacket and trousers. They wondered who he had been and how he had died. The archaeologists started calling him "Gustafsson" (although he later received the official designation "Helge", in the alphabetical ordering of the skeletons).

They carefully washed and brushed off the mud and grit, while others dismantled and moved away the gun carriage pinning him to the deck. They made drawings and took photographs, and a discussion began about how to recover his remains. Because there was more than bone preserved – they had found fingernails and hair, and could even glimpse his brain in the shattered skull – and because he was still wearing his clothes, which might have smaller objects in the pockets, they decided to lift the skeleton together with the surrounding mud so that the find could be documented and cleaned under laboratory conditions. Slowly, respectfully, they slid thin sheets of metal under the body and lifted him, not as a collection of artefacts, but as a person. Over three centuries too late, he finally left the ship.

A MILESTONE IN ARCHAEOLOGY

Archaeology in general, and the *Vasa* project in particular, is a dialogue between the living and the dead. It is a way to reach across centuries, to people who lived in other times and places, and ask them about their lives. The building of a new type of warship in Stockholm in the 1620s involved thousands of people, from loggers in Poland to weavers in Holland, tar burners in Finland and miners in the Great Copper Mountain of Sweden. An international workforce of hundreds

cut and assembled the ship at the Crown's navy yard, while others cast the guns, sewed the sails and laid the rope. It was a major social, political and economic event, and all of the people involved left some trace in the ship. Those traces are there for us to read, an opening into a lost world.

Although never entirely forgotten, the wreck gradually disappeared from the conscious memory of the majority of Swedes until the 1950s, when two civilian employees of the Swedish Navy relocated the wreck. Anders Franzén, a fuels

The salvage of *Vasa* was a manly business. The "Tunnel Gang" pose for a publicity photo as if in a scene from a Tintin adventure. The dive boss, Per Edvin Fälting, sits in front, surrounded by Sven-Olof Nyberg, Stig Friberg, Lennart Carlbom and Ragnar Jansson.

A SHIP'S TALE   15

A gilded lion from the stern, a symbol of royal power, is carried ashore by Per Edvin Fälting (right). Because it had lain in the mud, it was exceptionally well preserved. Much of the gilding survives today, although covered by the preservative chemicals used to treat the ship.

engineer, and Per Edvin Fälting, a salvage diver, brought the ship back into the public eye in the autumn of 1956. What followed could only have happened when it did, as the post-war economic boom made grand projects seem possible and tangible memories of Sweden's history desirable. Other big wrecks had been found before, but it was normal practice to dynamite them to recover the valuable bronze guns and waterlogged "black" oak. Instead, Franzén had a vision of an intact ship as "a cultural monument of international standing [and] an effective but above all necessary reminder of Sweden's Great Power Period and naval traditions". Crucially, he also had the ability to recruit others to his vision. He persuaded the navy, the National Maritime Museum, the National Heritage Board and the Neptune Salvage Company to collaborate in an audacious plan, to raise the ship intact from her resting place on the bottom of the harbour, conserve and restore her, and create a museum that would attract visitors from all over the world.

It is now fifty years since *Vasa* returned to the surface, and his vision has proven to have been the right choice. Over a million people annually visit the

Vasa Museum and take away an image of the past that is unique. For a brief moment, they can experience an earlier age and meet those who built and sailed this ship. A complete failure as a physical warship, *Vasa* has outlived all of her contemporaries to become an unparalleled success as a metaphysical warship. This is her story, and the story of the dialogue between shipwrights and sailors of the 17th century, divers and archaeologists of the 20th century, and researchers and visitors of centuries to come.

King Gustaf VI Adolf (left; r. 1950–1973) was a keen amateur archaeologist and visited the excavation several times. Here he is shown the current state of the dig by Per Lundström (right), the archaeological director.

An age of faith and violence

1600

By any standard, the 17th century was an exceptionally violent age. Countries and empires under severe economic and ideological stress ground against each other in an increasingly crowded Europe, striking sparks. They exported their conflicts by sea to infant colonial empires and global trading networks. They fought over the exotic riches of the orient and the more mundane bulk goods of the Baltic, over political independence and dynastic inheritance, over faith and the nature of God's grace. Using new technology on land and sea and the new wealth generated by globalized commerce, they expanded the scale of warfare and its attendant devastation, spawning a desperate scramble for the resources to keep fighting. Armies grew from thousands of men to tens of thousands, kept under arms for years at a time. Increasing numbers of conscripted citizens formed new regiments, while the recruitment of the mercenaries who made up the bulk of forces intensified, as rulers sought new solutions to the demand for manpower. The new large armies denuded the countryside they passed through and emptied the treasuries of their employers. Navies ceased to be fleets assembled for a particular campaign and became permanent institutions, which consumed timber, metal, men and victuals year-round, whether they fought or not. There were great opportunities for the ambitious (and lucky), and great suffering for the ordinary farmer, fishermen or craftsman trying to feed a family.

Nearly every country was at some point at war with its neighbours, with the great powers of Spain, France and the Holy Roman Empire each attempting to be the leading force in Europe and the others forming temporary combinations to prevent any one country from achieving lasting dominance. In addition to this balance-of-power dance, a general alignment began to emerge as a result of the Protestant Reformation. The teachings of Martin Luther, John Calvin and others had taken hold in northern Europe, primarily in the Germanic and Scandinavian countries, while the south remained Catholic. Differences of confession led to political conflict, as Protestant princes were happy to be free of allegiance to

Sweden reinvented a significant part of its glorious past in the design of Gustav Vasa's grave monument in Uppsala cathedral in 1583. It is decorated with the coats of arms of Sweden's different regions, in the style of a European feudal monarch.

Since at least Viking times, the area around Lake Mälaren in central Sweden has had a strong political, cultural and commercial influence flowing out through the Stockholm archipelago to the Finnish archipelago and eastward to the end of the Gulf of Finland. In *Vasa*'s time, the main Swedish political axis still followed an east-west line, although the young King Gustav II Adolf would try to change this.

The leading economic power in northern Europe was the United Provinces (now the Netherlands), fighting for its independence from Habsburg Spain for the first half of the 17th century. Maurice of Nassau, Prince of Orange, was the nominal head of government at a key period and a military innovator. His tactical theories on small, mobile units working in coordination were taken up and further developed by Sweden under Gustav Adolf.

Rome, but Catholic sovereigns felt obligated to bring heretic subjects back to the true faith. Many of the wars of the 17th century thus took on religious trappings, either as cause or window dressing.

In northern Europe, the Protestant Dutch had been at war with Catholic Spain since the 1560s, attempting to extract themselves from the rule of the Spanish branch of the Habsburg empire. The German Habsburgs were struggling with their own independence movement, as the hundreds of kings, princes, dukes, counts and knights who governed the small and large fiefdoms that made up the Holy Roman Empire chafed at imperial rule, with the Protestant princes of northern Germany some of the most vocal in their criticism of the Catholic emperor. In the Baltic, the fate of the wealthy grand duchy of Lithuania concerned all of its neighbours, who fought to decide if it would be part of Poland, Russia, or Sweden. At the same time, Denmark and Sweden struggled over dominance in the region, especially at sea. England and Scotland (ruled by the same king after 1603) meddled in most of these wars in some form and were a source of mercenaries, but eventually succumbed to internal tensions, dissolving in civil war in the 1640s.

Many of these wars were essentially about resources, as ambitious princes and nobles jockeyed for advantage and tried to extract income from the trade passing through their lands or gain control of the trade passing through their neighbours' possessions. At the same time, the social and ideological stress brought about by the Protestant Reformation and Roman Catholic attempts to suppress or overturn it were reaching a peak. Debates became ever more shrill, the flames of fear and hatred fanned by priests and princes until ordinary people were willing to kill each other over what they believed.

Much of this tension came to a head after the Protestants of Bohemia refused to submit to the Catholic Habsburg emperor and rebelled in 1618. This local revolt ignited the great fire which many had expected for decades. A general war developed, drawing in virtually all of the German princes and their neighbours, eventually including France, Denmark and Sweden, before a general peace was agreed in 1648. This related series of conflicts, now called the Thirty Years War, was the bloodiest in human history before the 20th century. More than a million died and the rich farmland of central Germany, from Bavaria to the Baltic, was laid waste, wrecking the regional economy for two generations and creating a

general European economic crisis in the 1620s. The war rapidly moved past its religious and local origins, as the northern German princes saw an opportunity to increase their autonomy, and other Protestant rulers, particularly in Scandinavia, sought to help them or to exploit the situation for their own gain.

ON THE THRESHOLD OF POWER

Sweden in 1628 was a very different country than it is today, both larger and smaller. Finland was still a part of Sweden, but the south-western counties of Halland, Skåne and Blekinge as well as areas of the western fell country were still integral parts of the Danish-Norwegian kingdom. Wars in the early part of the 17th century had brought areas of the south-eastern Baltic coast, in modern Russia, Estonia, Latvia and Lithuania, under Swedish control. The total area was immense and rich in natural resources, primarily timber and metals, but thinly populated and underdeveloped compared to continental Europe. There were few settlements large enough to call towns, and the only one that could claim to be a city in European terms was Riga, recently conquered from Poland. Stockholm was a large town, with about ten thousand inhabitants, but it was still built mostly of wood.

The population in Sweden and Finland was overwhelmingly made up of free peasants, who owned most of the arable land. There was a small noble class, who were on average quite poor by continental standards. The economy was

The road to greatness was not always smooth. One of Sweden's greatest defeats came in 1605 at Kirkholm, near Riga. 9,000 Swedish soldiers died before Poland's magnificent cavalry. The event was later memorialized in this Polish painting from the 1920s, after Poland regained its independence and could point with pride to its own Golden Age in the early 17th century.

Swedes and Danes viewed each other with suspicion, even if they were temporarily allied in the 1620s against a common enemy, the German emperor. The Danish fleet was a match for the Swedish, and had a better reputation for seamanship.

The Bourse in Amsterdam was founded for the Dutch East India Company, but eventually became the central clearing house for Baltic goods as well.

NORWAY (DENMARK)

Gothenburg

SWEDEN

Kalm

DENMARK

Copenhagen

SKÅNE (DENMARK)

SCHLESWIG-
HOLSTEIN
Lübeck
Wismar
Stralsund
Rostock
Greifswald
POMMER

MECKLENBURG

Hamburg

OLDEN-
BURG

Stettin (Sczecin)

THE UNITED PROVINCES

LÜNEBURG

BRANDENBURG

• Berlin

•Amsterdam

SPANISH NETHERLANDS

SAXONY

Dresden

HOLY ROMAN EMPIRE

Åbo
(Turku)

SWEDEN

Stockholm

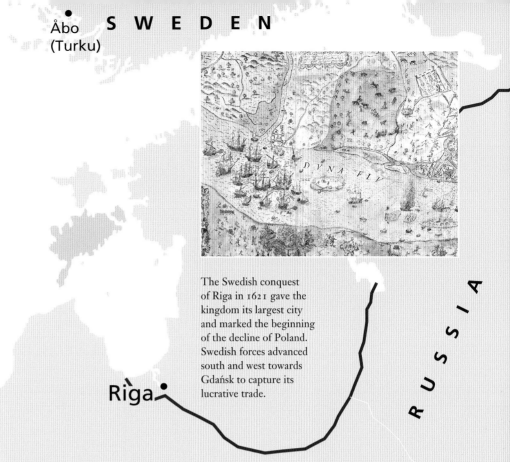

The Swedish conquest of Riga in 1621 gave the kingdom its largest city and marked the beginning of the decline of Poland. Swedish forces advanced south and west towards Gdańsk to capture its lucrative trade.

Västervik

GOTLAND
(DENMARK)

Riga

RUSSIA

Memel

Königsberg (Kaliningrad)

POLAND

Danzig
(Gdańsk)

1628

THE GEOGRAPHICAL CONTEXT

Sweden was the rising power in the Baltic, first at the expense of Russia (which had already abandoned Estonia in the 1580s) and then Poland (which lost Riga and the Livonian lands to the north in 1621). Sweden regarded Denmark as its main rival throughout the Great Power Period. Swedish expansion depended on Dutch capital and expertise to a large degree, and the Dutch saw Sweden as a useful counterweight to Danish control of Öresund.

The central figure in Dutch economical and political life was the self-confident merchant. Merchants controlled the government and used it aggressively to promote trade and protect their business interests.

agricultural but without well-developed markets. There was little manufacturing of significance, and most of the small luxuries taken for granted in England or Germany had to be imported. Only the metals industry put the country on the international economic map – the mine at Falun produced most of the copper used in Europe, and the mines north of Lake Mälaren exported significant quantities of high-grade iron. Roads were few, and communication over any distance depended on coastal and inland shipping.

Swedish vessels were only a tiny fraction of the shipping on one of the world's busiest waterways, between the Baltic and the North Sea. Baltic timber, grain and fish were sold to the growing cities of the more developed countries to the west, which were desperate for food and building materials. Ships entering the Baltic brought salt, wine, and manufactured goods, especially woollen cloth. This bulk commerce involved thousands of ships and was the most lucrative in northern Europe, much more so than the trade in Far Eastern spices. Dutch merchants had wrested control of this trade from the Hanseatic League in the 15th century, and it was the foundation on which the prosperity of the Dutch Golden Age was built. The king of Denmark also profited from Dutch success. He controlled both sides of the narrow Öresund strait leading into the Baltic, giving him the chance to levy a toll on every ship entering and leaving. The few Swedish vessels trading outside the Baltic were normally exempt from the toll, and this exemption was usually one of the privileges confirmed in treaties ending the frequent wars between Sweden and Denmark.

From 1397, Sweden had been part of the Kalmar Union, under which the three Nordic countries were ruled by the same dynasty, the Oldenburgs of Denmark. In the 1520s, a Swedish nobleman, Gustav Eriksson, led a rebellion against Christian II of Denmark and established Sweden as an independent kingdom. He became king in his own right in 1523, as Gustav I, although he is usually known to modern Swedes as Gustav Vasa. He and his successors did not rule as absolute monarchs, but with the consent of the nobility. Relations between the king and the nobles varied, but under Gustav I's successors, there was a broadly successful effort to recruit the nobles to a subservient role to the Crown. They were persuaded to invest their prestige and resources in the Crown's policies as the best chance for better access to power and wealth, and by the end of the 17th century, they had become an effective professional class of government administrators and military leaders. The king carried a great deal of personal power and influence, and until the 1610s, was the effective administrative head of all branches of government. Even after the king delegated power to a departmental bureaucracy, he was still the initiator of foreign and domestic policy.

Swedish kings believed themselves to be surrounded by their enemies. To the east, Russia desired better access to the Baltic. To the west, Denmark had been Sweden's traditional rival for Baltic power since the end of the Middle Ages. To the south, Poland presented not only a large continental state with a wealthy noble class, but also a dynastic challenge for Gustav's heirs. Since the outbreak of the Thirty Years War, the Catholic Hapsburg emperor had also begun to take an interest in Baltic affairs. All could see the value of the great torrent of goods sailing westward and could imagine the revenue it might provide to the country which could control it.

Since the mid-16th century, Sweden had frequently been at war with her neighbours over territory, access to the sea, and control of Baltic trade. From the late 16th century, Sweden adopted a policy of forward defence, moving the borders east and south to create buffer zones on enemy territory. This meant the conquest of new lands, primarily at the expense of Russia and Poland, bringing Sweden closer to the sources of the bulk commodities on which Baltic trade was based.

GUSTAV ADOLF AND THE POLISH WAR

Gustav II Adolf (1594–1632) came to the throne in 1611 as a teenager. He inherited three wars, with Denmark, Russia and Poland (the last more of a cold war in the 1610s). He lost the first, and as a consequence Sweden was forced to pay an extraordinary ransom for the return of Älvsborg, Sweden's only fortress on the west coast. He won the negotiations at the end of the second war, and with the help of a long period of internal instability in Muscovy, effectively shut Russia out of the Baltic for most of the 17th century. In the breathing space afterward, he set about winning the confidence of the nobles, recruiting them to his vision of a greater Sweden, and reorganizing the Swedish state and economy to allow the country to field the most feared army in Europe.

He was a devout Christian and a staunch Lutheran, suspicious of Catholic efforts to influence events in his kingdom. He saw himself as a great Renaissance

In Dutch eyes, Sweden was primarily a source of high-quality raw materials, particularly metals, and a small market for imported manufactured goods. It was also a useful ally against the Danes in keeping the Baltic open. Once Sweden became the dominant power in the Baltic, Dutch political assistance shifted to Denmark.

Gustav Adolf was wounded in the neck as he led the Swedish army against the Poles at Dirschau (modern Tczew) in August 1627. He ordered his bloody clothes preserved, forming the core of the Royal Armouries collection.

prince and Sweden as the natural leader of the Baltic region. For him, an expanded and dominant Sweden was the best guarantee of political stability and religious freedom, by which he meant a Protestant Christianity unthreatened by the Catholic Counter-Reformation. He eventually came to see himself and Sweden as players on an even greater stage, as the saviours of the beleaguered Protestants of northern Germany. In 1630 he took the country into the main theatre of the Thirty Years War in Germany, where his success on the battlefield established an international reputation for bravery and tactical brilliance. Swedish intervention in the war would eventually prove decisive, but he did not see the result of his decision, as he died at the battle of Lützen in 1632. Decisions he made in the last years of his life hint that he had even greater ambitions, perhaps to challenge the Spanish on the Atlantic or to set himself on the Holy Roman Emperor's throne. Gustav Adolf is often seen as the king who set Sweden on the path to imperial power and expanded international influence. The century beginning with his reign is often called the Great Power period or the Age of Greatness by Swedish historians.

For most of the 1620s, Gustav Adolf was at war with his first cousin, King Sigismund (Zygmunt) III of Poland. Sigismund had succeeded to the Swedish throne in 1592, but had already been elected king of Poland in 1587. The divided responsibility of two kingdoms was not popular with the nobles of either country. Moreover, Sigismund had been raised in the Catholic doctrine and there were fears in Sweden that he might bring the Counter-Reformation to Scandinavia. A group of Swedish nobles, led by Sigismund's uncle Karl, the third son of Gustav I, revolted and deposed Sigismund in Sweden in 1599 (he remained king of Poland until his death in 1632). Karl became regent until 1604, when he was proclaimed as King Karl IX and his son, Gustav Adolf, became the crown prince.

Sigismund never gave up his claim to the Swedish throne and schemed to unseat Karl and then Gustav Adolf. Eventually, Gustav Adolf tired of the threat. He could also see the potential revenue which the Polish bulk-goods trade represented, and a strong Poland threatened Swedish interests in Livonia (approximately modern Latvia) and Estonia. In 1621, claiming that he was fighting to assure peace and defend the Protestant faith against the threat of a Polish invasion, Gustav Adolf attacked Sigismund from the northeast, seizing the important Polish port of Riga and the surrounding Livonian territory. Riga was the second busiest port in the eastern Baltic, after Danzig/Gdańsk, and taxation of its trade provided a significant source of revenue to finance further expansion. The gains were less significant after 1621, despite campaigns almost every year, but Sweden retained Livonia and its port of Riga under the terms of the truce of 1629.

Sigismund (Zygmunt III in Poland) was king of Poland 1587–1632, the country's Golden Age, but could not forget that he had also been king of Sweden (1592–1599).

CREATING NAVAL POWER

One of Gustav Eriksson's early actions in his revolt in the 1520s was to create a navy, a small fleet of ships purchased from the German town of Lübeck. The fleet proved a decisive element in the capture of the important coastal towns and the lifting of the Danish siege of Stockholm. Gustav understood the need for maintaining supply and communication links by sea in a country without roads, and after he became king, he established a permanent navy. This was fairly advanced thinking, as it was not yet clear that a standing fleet of specialized warships was the optimum solution for naval power. Eventually it became the most common choice among large countries with a stable tax base, but Gustav I's initiative in a relatively poor country suggests how important he believed the navy was. In a kingdom of far-flung settlements largely connected by water, a navy offered flexibility and speed of movement. This was essential in what is now called the "projection of power" role of government, moving military force to a

An important component in the creation of a strong Swedish state was the building of a fortified city in the Dutch style on the west coast, Göteborg (Gothenburg). Gustav Adolf's fortifications were later expanded, with additions such as this tower, called Kronan, in the 1680s.

focus point ahead of the opposition. A navy was also a prerequisite for gaining control of Baltic trade.

He established a central maintenance facility for the fleet, the navy yard in Stockholm, but contracted for the construction of new ships at various shipyards throughout the kingdom. These tended to be located near major stands of timber, and so reduced the transport costs for raw materials. Ships were also purchased, but he realized that hiring vessels from a major commercial power like Lübeck risked dependence on a potential enemy.

Gustav I's sons continued to support and develop the navy. A number of large ships were built under Erik XIV (reigned 1560–1568), new bronze guns were cast and crews recruited. This investment in naval power proved essential in the Nordic Seven Years War of 1563–1570 with Denmark and Lübeck. Although it ended in a nominal Danish victory, Sweden was left in command of the Baltic. The naval battles in this war were some of the first to be decided by the use of cannon instead of hand-to-hand combat, and the first in the Baltic to use warships as weapons rather than troop carriers.

Gustav Adolf inherited a substantial and well-maintained navy from his father, but like the rest of the state's activities, it was managed by a small group of officials under the direct supervision of the king. Gustav Adolf believed that his highest calling as king was leadership in war, not administration, but he understood the need for a more rational system of government, if only to provide support for his war aims. He was fortunate in being able to recruit the leader of the nobles who had opposed Karl IX's autocratic rule, Axel Oxenstierna. He was one of the most able administrators of the 17th century, and his talent and interests complemented the king's vision and ambition.

In the late 1610s, Gustav Adolf and Oxenstierna began the process of creating official government departments (*kollegium*) staffed by professional bureaucrats to whom the king delegated his authority. This system was not formalized until 1634, but one of the first five departments to be organized was the navy. The king's authority was delegated to the Admiral of the Realm (*riksamiral*), an office held from 1620 by Karl Karlsson Gyllenhielm, Gustav Adolf's older half-brother. He was assisted by vice admirals as needed. In practice he delegated most of his real administrative responsibility to Vice Admiral Klas Larsson Fleming.

One of the most important practical consequences of their reforms was

the consolidation of the process of building new ships for the navy. Instead of contracting with yards all over Sweden, one ship at a time, new construction was concentrated into just two main centres, the navy yard in Stockholm and a private yard in the town of Västervik, on the Baltic coast to the south, with the occasional ship built in Göteborg. At the same time, they transferred the burden of administering the construction process to private entrepreneurs by leasing the navy yard and related facilities on a multi-year basis. Many of these entrepreneurs were foreigners, part of a wave of immigrant expertise and capital recruited to jump-start Swedish production of armaments and materiel.

The expensive war with Poland was financed partly by tariffs on Gdańsk's maritime trade, which was diverted to a Swedish toll station established at Pillau (modern Baltijsk in Russia). Here, Swedish warships ride at anchor in the Pillau roads.

THE SWEDISH FLEET IN THE 1620S

When Sweden captured Riga in 1621, the Swedish navy comprised about 30 warships supplemented by a number of smaller vessels, such as pinnaces (smaller full-rigged ships) and galleys (oared vessels). The typical warship carried a single full deck of cannon, with most of the larger ships armed mainly with 12-pounders (a cannon firing a solid iron ball weighing 12 pounds, about 5 kg). The navy's

AN AGE OF FAITH AND VIOLENCE   29

Denmark was the most capable seafaring nation in the Baltic. The Danish king established a powerful new naval base in Copenhagen in 1604, where he could equip and maintain the entire fleet behind massive walls and defensive outworks.

primary tasks were guarding convoys of troops and supplies to the war theatre, carrying dispatches, and blockading enemy ports.

Within the Baltic, the main rival for naval power was Denmark, with a navy of approximately the same size as Sweden's. The king of Denmark, Christian IV (r. 1598–1648), took an active interest in his navy and had the funding to hire foreign expertise, build ships and cast cannon. He was as ambitious as Gustav Adolf and could not afford to let Swedish naval power exceed his own. Denmark had the advantage of lying farther south, so Danish harbours were ice-free for a longer period and the fleet could put to sea earlier in the year, an annual worry for Sweden.

Sigismund hoped to counter Sweden's naval advantage and began buying, commandeering and constructing ships. At its high point in the late 1620s, the new Polish navy consisted of twelve ships, although no more than eight were ever operational at any one time. They rarely left harbour, but did score a signal victory against the Swedish blockading squadron at the Battle of Oliwa in 1627, the only fleet action of any size in all of Gustav Adolf's reign. At the end of the war with Sweden, Sigismund transferred his ships to the German emperor.

The emperor desired an Imperial presence on the Baltic, which might help Spanish merchants to make inroads on Dutch trade and thus weaken the Dutch in their war against Spain. He created a new office, General of the Baltic and Ocean Seas, for his influential commander, Albrecht von Wallenstein. Wallenstein began to assemble a fleet in Wismar, one of the few German Baltic ports held by Imperial forces. In the event, the only ships acquired were Sigismund's former navy, which were lost to Sweden with the capture of Wismar in 1632.

Gustav Adolf could see that it would take increased effort to maintain control of his route to the Polish coast, and that his enemies in the Baltic were increasing. During the 1620s, he embarked on a major programme of expansion. He ordered more ships of the existing types and more bronze cannon. But more ships alone would not suffice, and he initiated the construction of a new type of ship, with two full gundecks and armed with much heavier guns. These ships would be able to face any other single ship then sailing in the Baltic and could act as the flagships of sizable squadrons. Other countries were experimenting with such vessels, and the king may have been prompted by reports that Denmark's Christian IV was building ships with two gundecks. He may also have been looking beyond the Baltic to the North Sea, where the Spanish and Dutch navies dwarfed his own. As early as 1619 Gustav Adolf contracted with Dutch entrepreneurs in Västervik for a new, larger ship, named *Äpplet* (Apple), but the vessel was a failure and was sold back to its builders. By 1624, he was ready to try again. He ordered two ships of a new class, armed with a new type of artillery, to be built in the Stockholm navy yard.

Denmark's greatest 17th-century king, despite many military defeats, was Christian IV, whose "C4" cartouche marks many of Denmark's most imposing buildings. Like contemporary Swedes, he built in a Dutch Renaissance style.

AN AGE OF FAITH AND VIOLENCE

Timber, iron and men

1627

Margareta Nilsdotter's life was falling apart around her. Two years ago, she was the wife of a successful shipwright who had just taken over the navy yard and had a fat contract to build four new warships. The estates she managed had been prospering. Her brother-in-law Arendt had been travelling the Baltic and North Seas, buying the raw materials for the new ships. Her husband had been corresponding with the king, discussing the proportions for a new class of warships. The loss of ten ships that autumn meant that there would be even more work, building new ships to replace the losses. The future had looked bright. Then it had all come crashing down.

The first of the four new ships was behind schedule. They had launched it in the spring, but it had not been finished in time for service. Word had just arrived that two ships of the blockading squadron off Danzig (Gdańsk) had been lost to the Poles – might the new ship have made the difference between victory and defeat? The new copper currency had collapsed and everything was more expensive, but the contract was for a fixed price so they would lose thousands of dalers. The Dutch carpenters had gone on strike, refusing to be paid in the worthless copper coins and demanding silver. The Crown was not keeping up with the payments promised in the contract, so there was no cash to buy materials or pay the workers. To raise some money and keep the work moving, Arendt had sold some of the timber purchased for the next ship, but the Crown saw this as embezzling. He had been tried and only just escaped punishment. The king was increasingly impatient with progress on the ship.

And she had to handle it alone. Her husband Henrik, the master shipwright and manager of the navy yard, had died in the spring. She was left with his half of the contract and the responsibility of completing it. She had plenty of experience with bookkeeping and the management of rural estates, but knew nothing about shipbuilding. Arendt tried to help, but he had to keep travelling, buying materials and recruiting labour. Vice Admiral Fleming was hinting that the navy might

A trained eye was needed to select trees suited to the many different curved timbers needed in a ship's hull. Trees were felled and rough sawn in the forest to patterns sent out from Stockholm before being shipped back to the navy yard.

Oaks used in the construction of *Vasa* were felled at many locations around the Baltic, such as the estate Strömserum north of Kalmar. Oaks still grow here in a well-preserved cultural landscape.

1627
THE STOCKHOLM NAVY YARD

The Hybertsson family were one of Sweden's largest employers, with over 300 men on the payroll in the 1620s. The men were organized into specialist shops or gangs under the leadership of senior journeymen or masters, who in turned reported to Master Henrik. This diorama in the *Vasa* Museum shows the yard in early spring, 1627, with *Vasa* (centre) nearing the point where the hull can be launched and her sister, *Äpplet* (left) shortly after construction has begun.

The navy's activities were directed from the royal palace, Tre Kronor, which was also the arsenal for storage of the navy's cannon.

Logs were converted into planks by the sawyers, who worked in pairs and were the second-largest group of craftsmen in the yard after the carpenters.

The shipwrights followed the northern Dutch construction method, in which the bottom planks were temporarily fastened together until the frames could be inserted.

The navy yard lay on its own island, connected to the mainland by a bridge but closed off by a red gatehouse at one end of the bridge. The island was called Skeppsholmen, but the area is now known as Blasieholmen.

Originally a loft for sewing sails, the Big House became the main administrative office for the navy yard under Master Henrik. Recent excavations have shown that the buildings in the yard were more substantial than in this diorama, with heavy stone foundations to support multiple stories.

The large anchor smithy had several forges, which were large, open hearths, and employed many men. The master anchorsmith was one of the most powerful figures in the yard and one of the best paid.

Vasa, soon ready for launching. The rest of the topsides will be planked first, but the upper deck, sterncastle and beakhead can be completed once the ship is afloat.

have to take over the running of the shipyard. And now, in the depths of winter, the king was coming to inspect his new ship.

A NEW CLASS OF WARSHIP

In the autumn of 1624, the king and the admiralty had begun negotiating a new contract for the maintenance of the fleet and the construction of new ships. Since 1620, the Crown had not managed the Stockholm navy yard directly, but contracted with private entrepreneurs for several years at a time for the navy's needs. Under the first contract, with Anton Monier, the Crown had continued to supply raw materials to the yard, but the new contract would be on a purely cash basis. This would be a major change in the government's fiscal policy and was only possible because it had also contracted out the collection of taxes to private individuals. These tax farmers advanced a fixed sum to the Crown and then had the right to collect the taxes from particular districts. The amount they could collect over the advance was profit. Both types of contract were called *arrende*. Tax farming allowed the Crown to predict its revenues with greater confidence and to deal in cash with suppliers, but the temptation to exploit the population was too great for some contractors to resist and abuse was widespread.

The new navy yard contract, to begin in January 1626, would run for four years and was agreed with two Dutch brothers, Henrik and Arendt Hybertsson. Henrik was a master shipwright and had been Monier's partner for the latter years of the first navy yard *arrende*. He had worked in Sweden for nearly 20 years, building warships for the navy. Arendt, who is usually "de Groot" in the Swedish sources (he may have affected the name, which was associated with a famous Dutch commercial family; in Dutch sources he is Arendt Huybertszon), was a merchant with a wide network of contacts throughout the Baltic and North Seas. They were originally from Rijswijk, near Den Haag in South Holland, and had emigrated to Sweden early in the 17th century. Together they made an effective team, with Master Henrik building ships and managing the shipyard while Arendt travelled, purchasing raw materials. Henrik's wife, Margareta, managed their rural estates, some purchased and some granted by the Crown as part of the payment for their services, while one of their sons kept the books for the shipyard. This sort of diversified family business was typical of the period.

The Crown and the Hybertsson brothers negotiated the specific terms of the

In the 1590s, before he moved to Sweden, Henrik Hybertsson lived in this block of buildings on Warmoesstraat in Amsterdam (this is the canal side). While there, he was known as a merchant rather than as a shipwright.

contract during the autumn, as they tried to calculate the cost of maintenance and new construction on a cash basis. Eventually, two slightly different versions of the final contract were signed, one by the admiralty in December 1624 and one by the king in January 1625. This later led to some confusion, and correspondence and payment records, as well as the finished ship, indicate that Master Henrik and the Crown were following different versions.

The brothers agreed to maintain the hulls of all of the navy's ships, at a fixed cost (rigging maintenance was agreed with a separate contractor). They would build four new ships, two larger and two smaller, and construct all of the gun carriages for the fleet's cannon. They would provide and pay carpenters to sail on the navy's ships and look after their maintenance on active service. They also agreed to complete the unfinished work of the previous *arrende*. In return, the Crown gave them control over the navy yard, plus several iron works located around Lake Mälaren. The total price for maintenance and new construction would be disbursed by the treasury in monthly installments. The Crown also exempted the brothers from import duties on timber and charcoal, and allowed them to cut a certain amount of timber in royal forests.

The contract included general specifications for the new ships, although the two contracts differed on the dimensions and price. No dimensions were given for the smaller ships; they were to be the same size as the ship *Gustavus*, which Henrik had built under the previous *arrende*. These should cost 15,000 or 16,000 dalers each. The larger ships were to be either 135 feet (40.1m) long from stem to sternpost (the earlier version) or 64 *aln* (a unit of two feet, thus 128 feet or 38.0m) long on the keel (the later version). The length on the keel was about four-fifths of the length from stem to sternpost, so the latter ship would be substantially longer than the former, about 160 feet (47.5m) instead of 135. In both cases the beam was specified as 17 *aln* (10.1m). These ships were to cost 40,000 or 42,000 dalers each; oddly, the longer version would be the cheaper.

The smaller ships were the typical vessels that had made up the bulk of the fleet since the beginning of the century, but the larger ships were something new. The cost indicates that these ships were not simply longer than the norm but were substantially bigger overall. How much bigger was shown the following year, when the king ordered the armament for the first of the ships, 72 new bronze 24-pounders. This was unprecedented firepower for a single vessel and

Viby estate, north of Stockholm, was a royal property granted to Master Henrik in 1605 in return for his service to the Crown. It was managed, together with other estate income, by Henrik's wife Margareta. It is now largely covered by housing developments, but the approach to the main house has the modern name Mäster Henriks allé.

TIMBER, IRON AND MEN 37

A SOURCE OF CONFUSION

The Hybertssons' contract for the maintenance of the fleet and new construction was the subject of much discussion in the autumn of 1624. Several working copies survive, as well as two different final versions. They differ in several key details, and it was never clear which one was to be followed.

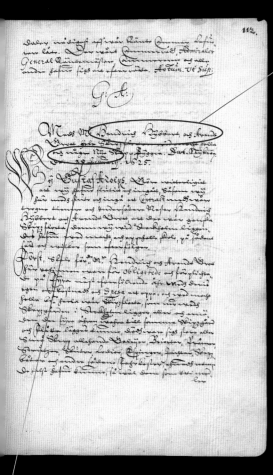

The two contracts involve different parties. Admiral of the Realm Karl Karlsson Gyllenhielm signed the version of 23 December, 1624, which names only Henrik Hybertsson. The version of 10 January, 1625, names both Hybertsson brothers and was signed by the king.

The state archives were badly damaged in the fire which destroyed the Tre Kronor palace in 1697 and many older documents were lost, but most that concern the construction and loss of *Vasa* survived.

The heading states the purpose of the contract: "With Mr. Hendrich Hybbert, and Arendt Grott, for the maintenance of the entire fleet and the construction of some new ships. Signed in Nyköping."

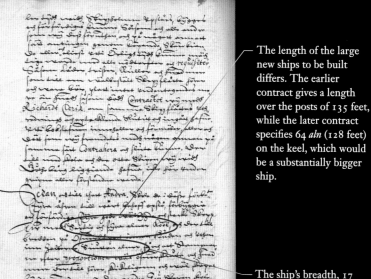

The length of the large new ships to be built differs. The earlier contract gives a length over the posts of 135 feet, while the later contract specifies 64 *aln* (128 feet) on the keel, which would be a substantially bigger ship.

The ship's breadth, 17 *aln*, is the same in both contracts.

The first of the two larger ships should be ready for service in 1626, the second in 1628.

The price differs slightly, with the earlier (smaller) ship costing 42,000 dalers and the later (larger) ship would require more than the single gundeck of earlier ships. When finished, these two ships would not be the largest ships in the Baltic, but they would be the most powerful warships in the world.

They would also be something of a risk. A few ships with two gundecks had been built elsewhere in Europe since the mid-16th century, such as Henry VIII's *Henry Grace à Dieu* in England and the *Galion du Guise* in Louis XIII's France, but not very many. Although ships with two gundecks and 60–80 guns would eventually be the standard battleships around which fleets were built, the process of finding the right combination of guns, shape and dimensions was still at an early stage. The problems of balancing strength, firepower, and seaworthiness were much more complex in such ships than they were in single-deckers, since the total deck loading was much higher and the lower gundeck had to be uncomfortably close to the water. Two gundecks made the ship inherently less stable than a single-decker. When the ship heeled under sail, the lower gunports on the lee side (away from the wind) could dip their sills into the water. If the ports were closed, this was not a problem, as the port lids had a double lip and were designed to seal, but if the ports were open, bad things could happen.

A contemporary picture of a French ship with two gundecks. Arendt de Groot claimed that the king was shown a similar ship as an example of what the Hybertsson brothers would build.

DISPUTED DIMENSIONS AND THE MYTH OF THE MEDDLING KING

By the autumn of 1625, Arendt had found and purchased sufficient timber to begin construction of a larger and a smaller ship, but there was insufficient labour to begin both simultaneously. Henrik still had to launch and complete his last ship from the old contract, a large single-decker. He contacted the king, through Vice Admiral Fleming, to confirm the order of construction agreed in the contract, which specified that the first ship to be built should be one of the larger.

Henrik supplied a list of dimensions of the ship about to be launched (not named in the document, but which can be identified as *Tre Kronor*) which he believed would be a good specification for the smaller ships. This was 108 feet (32.1m) on the keel, slightly different from the contract specification, and thus needed approval from the king. Fleming forwarded this list to Gustav Adolf, who was fighting in Poland.

While Fleming's letter was en route, disaster struck the Swedish navy. Ten ships, a third of the fleet, were lost in a storm off Domesnäs (Kolkas rags in modern

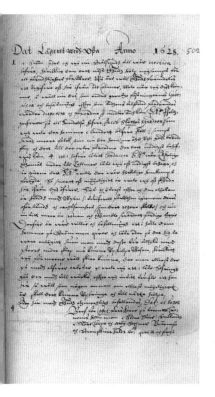

King Gustav II Adolf was in the field with his army in Poland on the far side of the Baltic in the spring of 1628, and sent a stream of letters back to Stockholm, demanding that *Vasa* be made ready and sent to her station.

Lithuania), near Riga. Gustav Adolf wanted to make up these losses as quickly as possible. He wrote back to Fleming with a new specification for a ship that would be between the larger and smaller in size, and ordered Master Henrik to build two of these. The king's specification was identical to the dimensions of *Tre Kronor*, except the length on the keel was extended by 12 feet, to 120 feet (35.6m). There were now three sets of dimensions under discussion, for ships 108 feet, 120 feet, or 128 feet long on the keel.

Increasingly tense letters between Stockholm and Poland followed. Master Henrik did not want to build the new, medium type. The timber for a larger and a smaller ship was already roughly cut. He could not use the smaller timbers to build the medium ship, and if he used the larger timbers, he would have to cut them down and thus lose money. The king insisted, the shipwright prevaricated, until finally, in the winter of 1626, Henrik was called to the office of the Chancellor of the Realm, Axel Oxenstierna, to explain himself. Master Henrik announced that he had already begun the first ship and it would be slightly smaller than the specification. Henrik Hybertsson may be one of the few people to have gotten the better of Gustav Adolf in a dispute.

From this exchange has grown one of the great myths of *Vasa*'s history. It has been claimed that one reason the ship sank is because the king, who was not a professional shipbuilder, interfered with the construction of the ship. Some have said that he ordered the ship made longer after construction had started, others that he ordered the addition of a second gundeck. In either case, the ship ended up poorly proportioned, top-heavy and thus unseaworthy, and it was the king's fault.

In fact, there is no evidence that this was the case. The discussion of dimensions occurred before construction began, and once the keel was laid, in late February or early March of 1626, there was no more argument about the new ship's size. The order for the ship's armament, in the summer of 1626, is clearly intended for a large ship with at least two full gundecks and had to have been made before construction of the hull was very far advanced. Most notably, the ship itself provides no evidence for substantial changes after construction began. There are several minor changes, including the moving of the mizzen mast and alterations in the internal arrangement of the cabins, but no discontinuity in the main structure. The ship that we see today in the *Vasa* Museum is more or less the ship the king ordered in the contract. It is very close to 128 Swedish feet long on the keel, and 160 feet over the stems, although wider than the 17 *aln* specified, closer to 38 feet to the inside of the planking.

FROM IDEA TO REALITY

The construction of a major warship began with the purchase of raw materials. Under earlier *arrende* contracts, the Crown had supplied timber, tar, iron and charcoal. Some of this came from royal estates and some had been paid to the Crown as in-kind taxes. The shift from in-kind taxation to tax farming and cash allowed the Crown to shift the burden of finding and collecting raw materials onto the entrepreneurs. The Hybertssons were paid a fixed sum in cash, and it was their responsibility to purchase materials and hire labour. Arendt de Groot spent most of 1625 travelling the Baltic and North Seas, visiting his contacts and making deals.

All of the materials needed for the hull were available within Sweden or its possessions, but in some cases it was economically advantageous to purchase abroad. In the case of timber, Arendt had a unique opportunity in the mid-1620s. The best timber for planking was the oak that grew in the forests of Poland and Livonia. This was normally bought at the source by Dutch merchants and shipped to the Amsterdam timber market, driving up the price in Baltic markets. All of this changed in the 1620s, when the Dutch were at war with Spain. Dutch shipping from the Baltic dropped precipitously, leaving Polish timber available at a competitive price. The Hybertssons' account books for 1625–1626 show that they took advantage of the situation and bought rough-sawn oak planks in large quantities in Riga and Königsberg (Kaliningrad), both in the Swedish-controlled area of Poland-Lithuania.

Arendt also bought planks in Amsterdam, which might seem strange. Much of the timber sold there came from Poland, so how could it be economically practical to ship timber out of the Baltic and then back again? Arendt's shipping contracts from Amsterdam survive, and show that he loaded oak planks for Stockholm in Amsterdam in 1626 and 1627, along with other supplies bound for the navy yard. Part of the explanation must lie in the Dutch control of the timber market and their efficient management of transport, but it may be that the oak bought in Amsterdam came from Germany.

The materials for the construction and maintenenance of the Swedish fleet came from all over northern Europe. The Hybertsson brothers owned a series of small ships over the years, which they used for transporting materials to Stockholm, but also paid other shippers to carry materials for them, costs noted carefully in their account books.

The curved timbers for the stem, frames and other elements which defined the shape of the hull were cut in Sweden. Trees had to be selected carefully, so that their natural curves would suit the eventual shape of the timber as closely as possible. For this reason, Master Henrik employed two other master shipwrights, Henrik "Hein" Jakobsson and Johan Isbrandtsson. Master Hein worked in Stockholm as the "yard master," leading the construction and maintenance work. Master Johan was the "forest master," travelling through the oak forests of central and southern Sweden in search of good timber. Master Henrik or Hein Jakobsson sent measurements or even patterns to him, so that he could select the right trees and roughly shape them in the forest. This minimized waste and shipping costs and assured maximum strength in the finished ship.

The purchase records show that in addition to the trees the Hybertssons were allowed to harvest from the king's forests, oaks were bought from a number of private forest owners. Most of the timber was cut either in the coastal region south of Stockholm between Kalmar and Norrköping, or in the forests around Lake Mälaren. Much of the timber was bought from nobles, such as Johan Skytte and Duke Johan Casimir, although the largest single seller was a woman, Mrs. Britta Posse of Ängsö, southeast of Västerås on the Mälaren shore.

The main structure of the hull was of oak, which was strong, durable and one of the few trees available both as long, straight lengths for planking and curved timbers for frames. Some of the decks were planked with pine, which was lighter, and other woods were chosen for particular applications. Alder was tolerant of wet conditions, so made good pump shafts. Ash was resistant to wear and shock, which was necessary for rigging hardware. Linden took fine detail well and was used for some sculptures.

Iron came from the mines north of Lake Mälaren and was shipped to Stockholm in the form of rough bars. These were then used to make the ship's hardware and fastenings at the navy yard, so large quantities of charcoal were needed to fuel the forges. Shipbuilding often takes the blame for the deforestation of Europe in the post-medieval period, but the iron and copper industries were the largest consumers of timber in Sweden. It may have taken over a thousand trees to make the frames and planking of *Vasa*, but thousands more were burned to make her bolts, anchors and nails.

Labour also had to be found. When the first *arrende* for the navy yard was signed in 1620, there were about 150 men employed there. Most came from Sweden and Finland. This was not enough to build the expanded navy the king wanted. Under Monier's *arrende*, Master Henrik probably had the responsibility of recruiting new, trained shipwrights. He found them in the land where he was trained, and by the mid-1620s there were over 300 men working at the navy yard. Most of the new recruits came from the Low Countries and worked only part of the year in Stockholm, returning home each winter.

We can see the international character of the workforce in their tools. Six carpenter's rules have been found in the ship, most in places that show that they were lost during the construction of the ship. Most are one-foot rules divided into inches, but no two are the same length. Some are divided into twelve inches, which is typical of the Swedish foot, although none are the standard length of 297mm. Others are divided into eleven inches, which is the type of foot used in Amsterdam and northern Holland, although only one is the regulation length of 281mm. The overall dimensions of the ship suggest that the Swedish standard foot was the "official" measure used during construction, but every carpenter

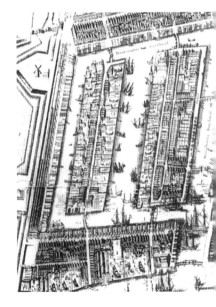

Shipyards in Holland operated on a proto-industrial scale. The materials and labour were well organized, so that ships could be produced in series, with deliveries taking months rather than years. Dutch shipwrights were in demand throughout northern Europe, and many were recruited to Denmark and Sweden.

TIMBER, IRON AND MEN

The only naval building surviving from Vasa's time may be the bakery (it may be slightly later), which lies on the opposite shore of Nybroviken, directly across from the building slips. Today it is the Music Museum.

Cost in dalers for the construction of *Vasa*'s hull:

Item	dalers
Salary to Dutch carpenters:	10 400
Swedish carpenters:	5 300
Sawyers and borers:	2 400
Sculptors and joiners:	4 600
All types of iron bolts:	8 700
Other iron fittings with anchors:	2 300
All types of nails and brads:	5 500
Oak planks, 1500 pieces at 3 dalers:	4 500
All types of oak timber:	9 600
Total:	53 300

Not included: shipping costs for the timber, arming cloths, and some fittings.

made his own rule. For measuring the dimensions of individual timbers, the small differences in the length of an inch were not significant.

We are fortunate that *Vasa* is the first Swedish ship for which an accounting of the actual build cost survives. It was compiled after the sinking and shows that the Hybertsson brothers had lost money on the contract. The total cost of the hull was 53,300 dalers, at least 27 per cent over budget. The increase was mostly likely due to inflation brought on by the collapse of the new copper coinage, introduced in 1624. Of the total, timber represents only 26 per cent of the cost. The iron, for anchors, bolts and nails, was 31 per cent. Labour was the largest single component, at 43 per cent. In overall terms, the cost was about 1000 times the annual salary of an ordinary sailor in the navy, but it is not possible to convert it into a number that makes sense in modern currency, since the relationship between wages and prices has changed so much since 1628.

The shipyard was one of the largest employers in Sweden. It was located on an island called Skeppsholmen (now the peninsula of Blasieholmen) to the northeast of the centre of the town. Raw materials arrived at the island from all over northern Europe, and left as finished ships. There were several building ways, a careening dock for doing maintenance work on the bottoms of floating ships, storage depots and specialized workshops for the construction of small boats, gun carriages and pumps and rigging hardware. There were two smithies, one for making anchors, bolts and other fittings and a smaller, specialist shop churning out the thousands of nails needed. The shipyard was large and important enough that its major buildings appear on the earliest map of Stockholm, from the mid-1620s.

The workforce was divided into groups based on their tasks. Sawyers converted rough timber into ship parts, while carpenters shaped and fitted the timbers together. The boat, gun carriage and turner's shops each had its own dedicated group of craftsmen, as did the smithies. Master Henrik presided over all of these and provided the key design information for each new ship.

The design and construction process was the same that Master Henrik, Hein Jakobsson and the Dutch shipwrights had learned during their apprenticeships in Holland. No drawings were required, so there are no lost drawings that would have shown the ship's form or dimensions. Instead, proportions derived from the basic dimensions of the ship were used to determine the main features of

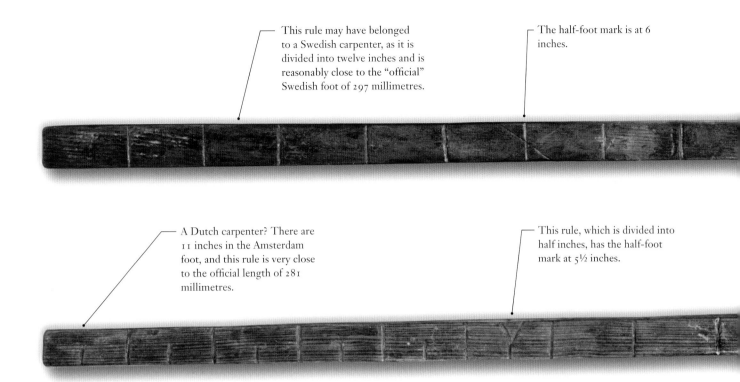

This rule may have belonged to a Swedish carpenter, as it is divided into twelve inches and is reasonably close to the "official" Swedish foot of 297 millimetres.

The half-foot mark is at 6 inches.

A Dutch carpenter? There are 11 inches in the Amsterdam foot, and this rule is very close to the official length of 281 millimetres.

This rule, which is divided into half inches, has the half-foot mark at 5½ inches.

the hull shape. The key dimension was probably the length of the keel, as Dutch shipbuilding treatises from the later 17th century show. Using proportional rules, the width of the ship, the rake of the stem and sternpost, the height of the decks, even the width and thickness of individual timbers could be worked out. For example, Cornelis van Yk, a professional shipwright in Holland, wrote in his treatise on shipbuilding (first published in 1697) that the keel's width and thickness should be 1/77 of the overall hull length. In *Vasa*'s case, this works out to about 24 inches, which is, in fact, the original width and thickness of the keel. If *Vasa* had started as a smaller ship, the keel would be proportionally smaller.

In most large-scale wooden shipbuilding since about 1500, the normal sequence of construction was to erect the keel, stem and sternpost as a backbone on which to build the rest of the ship. To this were fastened the frames, the main transverse structural elements. The shapes of these determined the form of the hull. The planks were then fastened to the outer surface of the frames with nails and treenails (wooden pegs), and a second layer of planking, called ceiling, was fastened to the inner surface of the frames. This was then reinforced by beams across the ship to maintain the hull's width and carry the decks. In warships, the heavy deck loads from the cannon were distributed down into the hull by an addition system of heavy reinforcing timbers, called riders.

The construction of *Vasa* was a multi-cultural project, as is shown by the carpenter's rules found in the ship – no two are alike!

TIMBER, IRON AND MEN   45

The Dutch were extraordinarily economical in their use of timber. Instead of discarding a plank with a knot, flaw or rotten spot in it, they would carefully cut out the bad wood and nail in a small graving piece. This type of thrifty repair is often called a Dutchman by English and American shipwrights.

In the Dutch tradition, dating back to at least Roman times, the finished structure is the same as the above, but the sequence of assembly was very different. After the keel and posts were set up, the bottom planking was laid up and temporarily fastened together with wooden blocks nailed across the seams. Once the bottom was complete, the blocks were gradually replaced with the frames and the nail holes filled with little wooden pegs, which can still be seen in the planking. After the bottom was framed, the frames of the sides were set up on the finished bottom, and the hull was completed in the conventional fashion, fastening the planks to the frames. Riders, if needed, were added to the finished structure.

This method had the advantages that it was very efficient in its use of wood and it did not require a complex geometrical or mathematical process for determining the shapes of the frames. It had the disadvantages that there was no way to evaluate the hull form in advance and no paper trail to allow a bureaucratic government to track or control the process. Later in the 17th century, this method was largely abandoned for big ships, partly because of the growth of naval bureaucracy and its demand for accountability and record-keeping.

The process could go very quickly in a well-organised and well-financed shipyard. Dutch commercial yards in this period could build and launch a large merchant ship in a single calendar year. *Vasa* took longer, partly due to the extra reinforcing timbers needed to carry the weight of the gundecks. After the keel was laid in the winter of 1626, the ship was probably not launched until the spring of 1627, about the time that Master Henrik died. He was already ill in the summer of 1626 and handed over the practical supervision of *Vasa*'s construction to Hein Jakobsson. At the launch, she was far from complete, but it was normal to finish the upper works and interior of the ship afloat. She lay at anchor for a year, while the carpentry was finished and the ship was rigged, and where Gustav Adolf inspected her in January, 1628. In the spring of that year, she was finally towed round to the royal palace, to load her armament and provisions in preparation for sailing.

Sometime in 1626 she received her name. It is first mentioned in the *arrende* for the maintenance of the fleet's rigging. She is first in the list of the navy's warships, as *Nya Wassan*, the New Vasa. She was the third ship to bear this name, which is the Swedish word for the Latin *fasces*, a bundle of sticks tied together.

In ancient Rome it was the symbol of judicial authority. This was the heraldic symbol of Gustav Adolf's family, although by the time *Vasa* was built it usually looked more like a sheaf of grain. In later centuries it came to be used as the name for the dynasty descended from Gustav I. Gustav Adolf had invested not only the kingdom's resources in this new ship, but his personal prestige. The ship would be a new milestone in his and the country's journey from the European backwoods to the forefront of the international stage.

Massive reinforcing timbers, called riders, help to carry and distribute the heavy weights of the gundecks. Down in the hold, the largest of these riders are 45–50 centimetres square, several metres long and held in place with dozens of iron bolts.

The Machine of War

1628

Medardus Gessus should never have accepted the order. It was not physically possible to cast so many large guns in so short a time. Seventy-two bronze demi-cannon, nearly 700 *skeppund* (90 tons) of copper! They were all to the same pattern, but no one had ever ordered so many guns in a single series before, or with such a short deadline. The patternmakers and foundrymen were working as fast as they could, but mistakes were being made. It was hard to get the bores centred in the moulds, and the clay moulds themselves were weak. A number had cracked when the molten bronze was poured in, some so badly that the breeches of the guns were distorted, with chunks of mould dross trapped in the metal. It was proving difficult to control the temperature of the moulds and metal as well, so some of the guns were porous near the muzzle and might not survive proofing, when they were test-fired with a double load. And he was still in the process of moving the foundry from its old location to Brunkeberg, north of the town centre, so tools and materials were still being sorted out.

The cannon was a new type, originally developed for the army as a lightweight siege gun. It fired a solid iron ball weighing 24 Swedish pounds (10.0 kg), and so was eventually known in service as a 24-pounder. The gun weighed about 9 *skeppund* (1200 kg), which was about half the weight of the old naval 24-pounder.

Even though the guns would all be made to a single design, each cannon required its own mould, which was destroyed in the casting process. The mould had to be made by hand, over a pattern that was also destroyed. The furnaces could only heat so much bronze at once, and the consumption of charcoal was astonishing. Once the guns were cast, the mould had to be broken off, the bore cleaned out, and the surface hammer-hardened and polished. The foundry was simply not big enough for the job, and the finishing work had to be contracted to another workshop.

Shoddy workmanship in one of *Vasa*'s surviving guns: the reinforcing rings should be evenly round, but are flattened and broken due to cracks in the mould during casting.

The mine at the Great Copper Mountain in Falun, the source of most of Europe's copper in the 17th century, as well as Sweden's primary source of foreign income. The copper mined here was the main component in *Vasa*'s bronze cannon. The mine's most prominent feature today is the "Great Pit," created by a massive cave-in in 1687.

Cannon founding in Sweden was developed largely by foreign capitalists, such as Louis De Geer, who made guns at Finspång (above). *Vasa*'s guns were cast at the Styckgjutargården in Stockholm. The ground plan survived, with new buildings (below), until the block was torn down in the 1960s.

In practice, it was most efficient to cast several guns at one time. It took many days of heating the furnaces to get up to 110 *skeppund* (15 tons) of metal to the right temperature, and it had to be poured in one smooth operation to avoid flaws. The moulds would be prepared and buried in the ground, with space for extra metal left at the top, necessary for maintaining pressure in the mould and minimizing shrinkage and cooling cracks. The raw material consisted mostly of new copper, delivered by the Copper Company, the semi-private Crown-owned business providing metal from the mines at Falun. Old guns and bells were also melted down and added, along with tin. The resulting alloy, called gunmetal, was 93–95 per cent copper with 4–5 per cent tin and 1–2 per cent zinc. The foundry could manage ten pours in a year, with a maximum of about ten guns in a single pour, but usually fewer. After pouring, it usually took another month or more of cleaning and polishing work before the gun could be test-fired. Those that did not pass would be remelted.

Gessus was not only under pressure because of the time and the size of the order. The king followed the activities of the gun foundry in Stockholm with an expert and interested eye. He had a direct hand in the development of this new gun. The king himself had proofed one batch of cannon in April 1627. Guns sometimes failed catastrophically in proofing, splitting or even shattering, and Gessus lived in terror that this would happen while Gustav Adolf was standing at the breech.

NEW GUNS FOR A NEW SHIP

Vasa's basic configuration, with two full gundecks and partial armament on the upper deck, was a relatively new idea for the time, but the armament was a uniquely bold and ambitious step in the development of naval warfare. Warships had carried gunpowder weapons for over two centuries by the time King Gustav II Adolf signed his contract with the Hybertsson brothers for the construction of the ship. At first, these had been small pieces, useful only at short range, but soon larger guns, capable of damaging a ship, were being carried in small numbers. There was a limit to how many such guns could fire over the railing, and the combined weight was dangerous so high above the water. At the end of the 15th century, shipwrights began placing the guns lower in the ship, firing through holes cut in the side of the hull. This reduced the stability penalty of carrying heavy cannon and made it possible for ships to carry more guns. Another century on, shipwrights were wrestling with the even more complex problem of multiple gundecks.

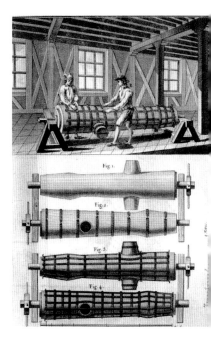

Making the mould for a cannon out of clay laid over a master pattern made of wood and rope. The mould was then reinforced with iron bands to take the pressure and weight of the bronze, which was poured into the mould after it was buried in the earth with the muzzle uppermost.

As these developments in ship design took place, the guns themselves were changing. The typical late medieval gun, made of forged iron staves and hoops like a long, skinny metal barrel, was rapidly replaced in the 16th century by solid, cast guns. These were in bronze at first, as the technology of casting large bronze objects was well understood from the making of bells and sculptures. As the more complicated process of casting iron was gradually tamed, this cheaper, stronger metal came to dominate.

Once gunports became common, shipboard heavy artillery fell into two main classes, based on configuration and size, although a wide variety of names was used for individual types. Large-bore short guns (cannon) were built for maximum hitting power at short range, while long guns of smaller bore (culverins) were used at longer range. The difference was due to the performance of early gunpowder, which burned slowly and inconsistently. The long barrel of the culverin allowed the powder to burn completely, developing maximum pressure and thus accelerating the ball to the highest possible speed. Improvements in the refinement of the

VASA'S GUNS

The main armament was a 24-pounder demi-cannon developed for the army as mobile siege artillery and first cast in 1620. It was shorter, with a thinner wall than traditional naval 24-pounders and became a standard gun for both the army and the navy by the 1630s. It was the Swedish answer to the demand for greater firepower in the limited space and carrying capacity of a warship.

A Dutch warship fires its guns in a broadside. Black powder created enormous clouds of smoke, which could obscure ships from one another if there was no wind.

The gun's elevation was set by a wooden bar under the breech and could not be adjusted.

The dolphins were located at the balance point and were used for lifting the gun.

The bronze tube weighed 1200–1300 kg, which was little more than half the weight of a traditional 24-pounder.

The length was 2.9 metres, considerably shorter than earlier guns.

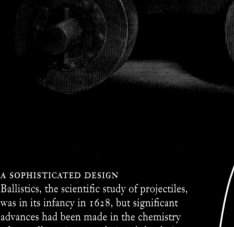

The gun carriage was made of ten pieces of oak plus a number of iron bolts and fittings, and weighed 300–400 kg. It was controlled by a heavy breeching rope and moved with tackles attached to hooks on the carriage.

Naval cannon normally had a smooth, flared muzzle so that the gun would not hang up on the edge of the gunport under recoil. *Vasa*'s guns have a muzzle more typical of the field artillery on which they were based.

A SOPHISTICATED DESIGN

Ballistics, the scientific study of projectiles, was in its infancy in 1628, but significant advances had been made in the chemistry of propellants (gunpowder) and the design of guns, which resulted in lighter, more efficient cannon.

shot acceleration

gas pressure

Acceleration levels off as pressure decreases. Muzzle velocity was relatively low by modern standards, although modern experiments have shown that speeds of over 500 metres/second are possible.

Seventeenth-century gunpowder burned very quickly compared to modern propellants. In *Vasa*'s guns, range was adjusted by using different sizes of powder charge. The largest normally used was 1/3 the weight of the ball, or 8 pounds (3.3 kg).

Maximum pressure is reached very quickly after ignition, so the tube had to be thickest near the breech.

The pressure falls off rapidly as the ball moves down the barrel, so the tube could be thinner.

As the ball exits the muzzle, it causes a shock wave. The swelled or reinforced muzzle kept the gun from splitting at this point.

raw materials and the manufacturing process produced faster burning, consistent powder. This powder could develop maximum pressure before the ball had travelled very far down the barrel, so a long barrel was not needed. The culverin class effectively disappeared by the end of the 16th century, and most naval guns after 1600 were variations on the cannon class. At the same time, the bewildering variety of size designations was abandoned in favour of nomenclature based on the weight of the ball fired. A Swedish "demi-cannon" (*halvkartog*) thus became a "24-pounder."

One of the limitations on the expansion of naval forces in Europe was the cost, in both materials and time, of casting guns. The process of manufacturing an object that required tons of molten bronze to be heated and poured in one continuous operation was resource intensive. This meant that arsenals were built slowly, over decades, and ships with large numbers of heavy guns were not very common. On the other hand, a well-made gun had a potential service life of over a century. The major disadvantage at the tactical level was that because each gun was an individual, it needed its own ammunition, carefully matched to the gun. The confusion and inefficiency this created in logistics and in battle can be imagined.

Gustav Adolf was himself a keen artillerist and took an active interest in the performance and development of cannon. As part of his effort to improve the effectiveness of the army, he had embarked on a program of standardization for everything from muskets to siege artillery to tactics. If all soldiers carried identical muskets, only one type of ammunition was needed and any soldier could use any other soldier's equipment. If artillery were reduced to just a few standard sizes, it would simplify the supply train, speed the loading process and make coordinated battery fire more effective.

If such measures could be applied to the army with success, then they should also work with the navy. Thus instead of the mixed bag of old and new guns that armed previous ships, the king ordered an entirely new armament of identical guns for *Vasa* in August 1626. The gun itself was still somewhat experimental. Throughout northern Europe in the 1620s, armies and navies were overhauling the design of their equipment to improve ballistic efficiency. A major effort was directed at making guns lighter. Lighter weapons were cheaper, easier to transport and allowed more guns to be carried by the same ship. Making them shorter,

Cannon had a long working life and the type of standardization seen in *Vasa* was not common until much later. When the Swedish ship *Kronan* was lost off Öland in 1676, it carried guns from the previous century and a half, such as this 16th-century culverin and several 24-pounders of the *Vasa* type.

THE MACHINE OF WAR

The Anglo-Dutch naval wars marked the development of more effective artillery tactics and the coordinated use of fleets and firepower. Here, the Dutch and English meet at the battle of Gabbard (1653), as depicted by Willem van de Velde the Elder, one of the pioneers of a new style of accurate, detailed marine art.

by focusing on cannon instead of culverins, was one way of reducing weight. Another was by making the barrel of the gun thinner. Better understanding of how the pressure of combustion built up in the barrel meant that metal could be placed where it was needed most and reduced in other areas, but in general a thinner-walled gun had to sacrifice range and hitting power. Less metal at the breech meant the gun could not take as high pressures as heavier guns, so a reduced powder charge had to be used. In the tactical environment of the time, this was not as big a disadvantage as it might seem.

The new demi-cannon ordered by Gustav Adolf was the Swedish response to the search for lightness. It was a gun with a relatively short bore, 17½ caliber (meaning the bore length was 17½ times the diameter) and a maximum wall thickness at the breech of only three-quarters of the bore diameter. By using two full decks of these guns, the king could quadruple the firepower over a single-decker armed with 12-pounders and simplify the supply and loading of ammunition. Firepower is calculated in broadside weight of metal, the weight of the iron thrown at the enemy by firing all of the guns on one side of the ship. *Vasa* would have a theoretical broadside over 800 pounds (330 kg), more than twice the broadside of almost all other ships of the time.

There seems to have been some awareness within the navy of both the potential danger of so many large guns and Gessus's difficulties. While *Vasa* was being built, a number of different armament plans were proposed. This was typical of the navy's planning process. A ship did not "own" its guns, but had guns issued to it for a particular mission or campaign season, and every winter the admiralty and the artillery board sat down to plan how to allocate the guns in the arsenal. The discussions were extensive, and had to accommodate the king's ever-changing plans of how the army and navy would be deployed in the coming year.

Of the surviving armament plans for *Vasa*, some were composed in the winter of 1626–1627, with the expectation that the ship would see service in 1627, and some in the following winter and spring. Most of these plans do not follow the king's vision of a unitary armament, but list 24-pounders on the lower gundeck only. The upper gundeck would be armed with conventional 12-pounders, and the upper deck with even lighter guns. This sort of graduated armament became the standard arrangement in the Swedish navy and everywhere else as the most logical compromise between stability and firepower. It is probably no coincidence that the gundecks, which were being built and fitted out while the first round of plans were being made, are constructed with a mixed armament in mind. The gunports on the upper gundeck are smaller than those on the lower gundeck, and the upper deck ports are even smaller. The total number of guns varies, but generally falls in the range of 60–72 guns. The total number of gunports actually built into the ship is 72, plus four small round ports in the transom for light swivel guns.

By the spring of 1628, the plans had reverted to a unitary armament of 24-pounders on the gundecks (a total of 56) and lighter guns (3-pounders and assault guns) on the upper deck. Even this proved too ambitious. Master Medardus simply could not cast an entire ship's armament in the time it took to build the ship, even when construction was behind schedule. The inability to arm the ship delayed its entry into service until late in the 1628 campaigning season. Eventually the king sent one of his artillery masters, Erik Jönsson, home from the Polish battlefields in the early summer to sort out the mess, get the ship armed and get her to sea.

The guns Gessus cast used the same basic pattern as the guns originally

Gustav Adolf was an innovator in the use of artillery on land as well. He standardized the army on just three sizes of gun and attached the lightest directly to individual infantry regiments, as at the conquest of Frankfurt an der Oder in 1631 (above).

THE MACHINE OF WAR 55

Gustav Adolf's initials GARS, (*Gustavus Adolphus Rex Sueciae*) appear above the coat of arms of Sweden on every gun.

Guns cast specifically for *Vasa* bear the date 1 6 2 6, when Medardus Gessus made the patterns for the decoration; those taken from army stocks are dated 1620.

Decorative bands of acanthus leaves and sea monsters appear only on the guns cast for *Vasa*.

The cascabel does not end in the button typical of later guns, but in a seahorse, which supports a small pan for the priming powder to fire the gun.

Although they differ in decorative detail, casting date and workmanship, *Vasa*'s three surviving guns are remarkably similar in basic dimensions and form, one of the first examples of series production of artillery. Red marks ridges made by cracks in the mould.

developed for the army, which was a plain weapon with only the royal coat of arms and the casting date as decoration. For the new *Vasa* guns, he added elaborate floral and figurative relief, but otherwise the guns were the same. The guns made specifically for *Vasa* have a casting date of 1626 on the breech, which is when he made the patterns (deliveries did not begin until 1627). When it proved impossible to complete the order in time, some of the plain, army-pattern guns were taken instead. These have a casting date of 1620. Even so, it was not possible to complete the ship's armament in 1628.

An inventory of guns on board the ship, made in the last weeks before *Vasa* sailed, shows a total of 64 bronze guns. Forty-six of these were the light pattern 24-pounder and two were old-fashioned naval 24-pounders, long and heavy. This left eight empty gunports on the gundecks, but the excavation of the interior produced all of the gun carriages needed. The carriage shop at the shipyard had completed its part of the job.

The upper deck was armed with a motley collection of antiques and cast-offs. Eight guns were old-style 3-pounders, spindly-looking culverins from the

16th century. Two were tiny 1-pounders, probably intended for the two small round ports on the quarter deck, although one did not yet have a carriage and was lashed to the deck when the ship sailed. The remainder were assault guns (Swedish *stormstycken*), short-barrelled, thin-walled guns for firing dedicated anti-personnel ammunition at the enemy's upperworks. These were of three different sizes, nominally sized to fire 20-pound, 48-pound or 82-pound shot, but in practice used for scrap metal or canister shot at very short range. The 20-pounder was a captured gun, one of a matched pair decorated with dragons, taken from the fortress of Prince Krzysztof Radziwiłł at Biržai (in modern Lithuania) in 1625.

The total weight of all of these guns was listed as 468½ *skeppund* (62.2 tons), which seems like a lot, but is typical for a ship of *Vasa*'s size. In most successful multi-decked warships, the weight of the guns was 5–7 percent of the total weight of the ship fully equipped – for *Vasa*, the guns are 5.1 per cent of the total displacement. The carriages, which weighed 300–400 kg each for the 24-pounders, added another 20 tons or so. Because not all of the planned guns were delivered, the ship's broadside was less than intended, but it still came to a nominal 753 pounds (312 kg), although the assault guns would never fire solid shot. This is the heaviest known broadside for any ship before the 1630s.

The guns were worth considerably more than the ship itself. The raw copper used to cast them was alone worth nearly 50,000 dalers, before the expenses of tin, charcoal and labour were added. Their value was not reduced by spending several decades under water after the ship sank, and eventually all but three of the guns were recovered from the wreck and sold.

NEW WEAPONS BUT OLD TACTICS

The ship and the armament may have been innovative in concept, but the tactics to use them effectively did not yet exist. The Swedish navy still operated on an older tactical model dating to the Middle Ages and pre-gunpowder weapons. Gustav Adolf was one of the first to break free of medieval concepts on land and coordinate the operation of infantry and artillery, so he may have been thinking in similar terms for the navy when he ordered the ship.

Admirals and captains still expected a battle to proceed in a traditional way. Captains received instructions in advance, giving them assignments and establishing the signals to be used, but once the battle began, each captain was on his

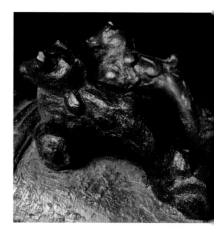

The lifting rings on naval guns were often shaped like sea creatures and came to be called dolphins. The dolphins on *Vasa*'s guns, even the ones originally made for the army, are in the form of seahorses.

1627

VASA AS A GUN PLATTFORM

Vasa represents one stage in the integration of ship and armament into a weapon system designed to inflict maximum damage on an enemy. Important steps had been taken in cannon design, gunpowder production, and ship construction, but much remained to be done. Battles were still decided by hand-to-hand combat, and the use of line tactics and concentrated broadsides was still decades away.

Marksmen with muskets could be placed in the tops at the mastheads.

The crew on the upper deck were exposed to enemy fire, but could fight back with a mixture of long-range 3-pounders and short range assault guns (*stormstycken*).

Quarter galleries were mostly decorative, but musketeers could be stationed here to fire along the ship's side.

Plan of the lower gundeck, fully armed with 30 guns. The guns covered a wide field of fire, like the spines of a hedgehog, which was typical for ships which fought as individuals.

Only the guns in the middle of the ship could be concentrated on a single target.

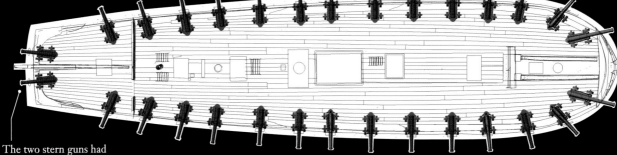

The two stern guns had not yet been delivered when the ship sailed.

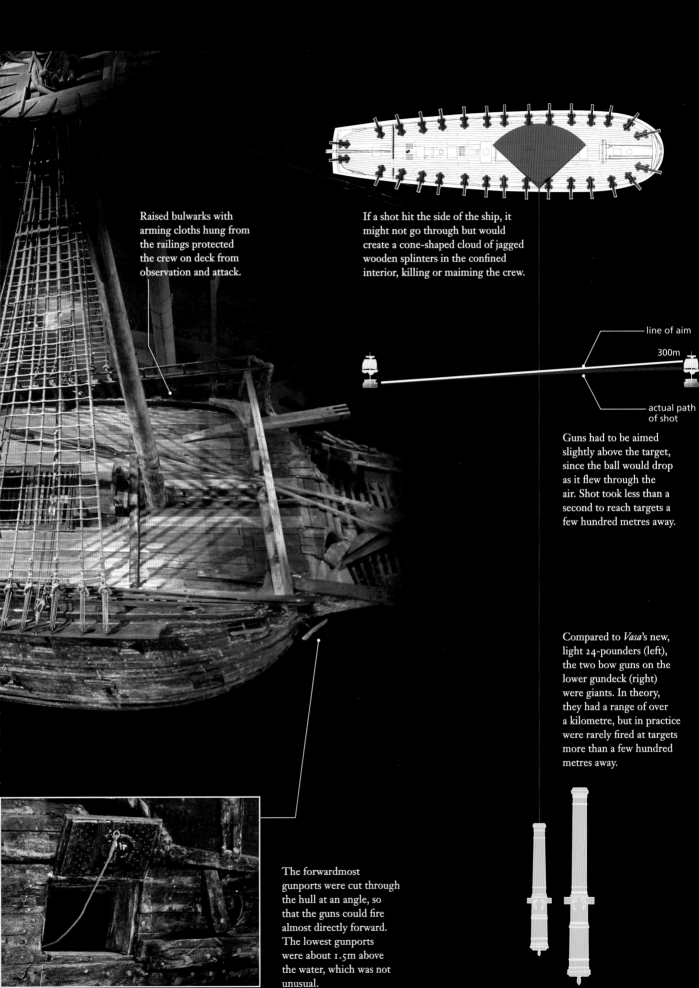

Raised bulwarks with arming cloths hung from the railings protected the crew on deck from observation and attack.

If a shot hit the side of the ship, it might not go through but would create a cone-shaped cloud of jagged wooden splinters in the confined interior, killing or maiming the crew.

line of aim
300m
actual path of shot

Guns had to be aimed slightly above the target, since the ball would drop as it flew through the air. Shot took less than a second to reach targets a few hundred metres away.

Compared to *Vasa*'s new, light 24-pounders (left), the two bow guns on the lower gundeck (right) were giants. In theory, they had a range of over a kilometre, but in practice were rarely fired at targets more than a few hundred metres away.

The forwardmost gunports were cut through the hull at an angle, so that the guns could fire almost directly forward. The lowest gunports were about 1.5m above the water, which was not unusual.

The battle of Oliwa, outside Gdańsk, in 1627 was the only fleet action during Gustav Adolf's reign. It was atypical in some ways, since it involved an attempt to escape a blockade, but was decided by boarding rather than cannon fire. It is recalled here in a diorama in the Central Maritime Museum of Poland.

own. He was expected to choose an enemy ship of equal size to his own vessel and attack it. What started as a fleet action quickly dissolved into a melée of individual ship-to-ship fights. If the sides were not evenly matched or the initial period of manoeuvre left one fleet in a dominant position, then several ships might gang up on a single enemy. It was not until Torstensson's War, between Sweden and Denmark in the 1640s, that Baltic admirals attempted to coordinate their ships' movements during the battle. In 1627, in the only significant naval battle of Gustav Adolf's reign, the small Polish navy escaped the blockade at Oliwa in Gdańsk Bay by going around one end of the Swedish squadron. This allowed all of the Polish ships to concentrate their fire on the last two ships in the line, *Tigern* (the Tiger) and *Solen* (the Sun).

The first stage in any action was a period of manoeuvring to achieve a better position. The ship or fleet to windward had the advantage, in being able to initiate or decline battle. With the sailing rig of the early 17th century, this was especially important, as ships had only a limited ability to sail to windward. It was common to reduce sail before engaging the enemy, in order to improve visibility and make the ship easier to handle. Ships could attempt to cripple an enemy with specialized ammunition for damaging the rigging, preventing escape or manoeuvre.

Once battle was joined, sinking the enemy was not the goal. Capturing the enemy ship and adding it to the navy was preferable. Tactics concentrated on demoralizing the enemy crew into surrendering before the ship was seriously damaged. It was a brutal business, in which wounding and maiming were more effective than killing. A dead man is just one man out of action, but a wounded man takes several healthy crewmen to take care of him. Specialized anti-personnel ammunition was used, although the traditional cannonball (called round shot) was effective on its own. Against the thick oak sides of well-built warships, round shot, even at short range, could not do much external damage and could not be guaranteed to go all the way through the side. The energy of the flying ball was still transferred to the ship's structure. It travelled through the wood and broke large, sharp splinters off the inside of the hull, throwing them around in the crowded, confined spaces of the gundecks, breaking bones and opening jagged wounds. Where the enemy crew were not protected by the ship's side, the cannon could be loaded with canister, wooden cylinders filled with musket balls or scrap metal, which cut a wide swath through tightly bunched men.

Round shot about 140mm in diameter and weighing 24 pounds (10.0 kg) were the main ammunition.

Spike shot was thought to cause more damage in close quarters but was not very effective.

Some of the 600 24-pound round shot found in one of the two shot lockers on board. One was in the same compartment with the galley amidships, this image is from the forward end of the hold.

Scissor shot were loaded closed and were supposed to open in flight, but were no more effective than spike shot.

Chain shot, two hemispheres connected by a chain, was the second most common. It was used to bring down rigging and tear holes in sails, so that an enemy could not manoeuvre or sail away.

The artillery duel between ships was only the preliminary action. It was expected that the battle would have to be decided by boarding and hand-to-hand combat between the crews. Ships did not carry large supplies of ammunition for the cannon, and the guns were only used at short range. When she sank, *Vasa* was carrying only enough powder to fire about 250 total rounds, or 5 rounds from each 24-pounder, although she had enough round shot on board for 12 rounds per gun. In later years, each gun would be equipped with wedges under the breech to adjust the elevation and thus the range, but on *Vasa* there was a simple bar across the back of the carriage to carry the gun at point-blank elevation. With the standard powder charge of a third of the weight of the ball, or 8 pounds (3.3 kg) for a 24-pounder, the effective range may have been no more than 400–500 metres. Even at this range, accuracy was poor, so it was normal to wait until the enemy was closer before firing.

It was not possible to bring all of the guns on one side to bear on a single target. Towards the ends of the ship, the ports were skewed fore or aft so that the guns pointed away from the ship at an angle rather than directly abeam. This increased the area covered by the guns, but it did not allow firepower to be concentrated.

HAND TO HAND

The primary weapon in the Swedish navy in the early 17th century was the crew, armed with hand weapons, and most battles were decided by hand-to-hand combat just as in storming a fortress on land. The standard firearm was the musket. Only a few were found on board, as the 300 soldiers assigned to *Vasa* were not yet embarked.

In the midst of battle (Öresund, 1658), the crews of small boats fire on each other with muskets.

In addition to a few military muskets, several hunting weapons were found on board *Vasa*. All of the iron has disappeared, leaving only the wooden stocks.

A wooden ramrod, for forcing the load down the barrel and compressing it, was carried in a groove under the barrel.

The musket was fired with a matchlock, the earliest form of mechanical firelock. A newer development, the snaplock (a type of flintlock) was already in use but the king preferred the simplicity and reliability of the older type.

Soldiers were armed with Swedish-made muskets of Dutch pattern, firing a lead ball 18mm in diameter (.71 calibre).

The long, heavy barrel was smooth on the inside, welded from a single sheet of iron rolled around a mandrel, and was loaded from the muzzle.

To fire, the musketeer had to blow on the lighted match, aim and pull the trigger, while holding the musket steady, often on a forked stick.

A premeasured charge, carried in wooden tubes on a bandolier, is tipped down the barrel.

After the ball and wad are added, they are driven home and compressed with the ramrod.

Loading and firing a matchlock was a complicated process which had to be learned systematically. Here we see 3 of over 40 steps in the most common manual of arms of the period.

In the Swedish navy of the 1620s, the ship was not yet the primary weapon, even if Gustav Adolf and others were beginning to think of how artillery might be better used. The crew were the decisive armament, and two-thirds of the crew proposed for *Vasa* were soldiers, two companies with a nominal strength of 300, commanded by their own officers and with no responsibility for sailing the ship. They served the guns in the early stage of the battle, under the direction of the master gunner and 20 gunners. Once the ship came alongside the enemy, they swarmed aboard the other ship, armed with boarding axes and pikes. Some would be stationed at the railings in the higher parts of the stern or in the rigging with muskets to fire into the decks of the enemy.

If the battle was about to be lost, the tactic of last resort was to blow up or burn the ship so that it would not fall into enemy hands. Charges were laid in advance so that they could be ignited quickly. At Oliwa in 1627, the commanders of both of the Swedish ships attacked by the Poles ordered the charges fired, but only one (*Solen*) succeeded. The man carrying the match to the other was cut down before he could get to the fuse, and his ship was captured and taken into the Polish navy.

Lathe-turned wooden canisters filled with musket balls or scrap metal were fired at exposed groups of enemy crew just before boarding parties were sent across.

In addition to the axes, pikes and muskets specified by the navy, some crewmen brought their own weapons. A handful of swords were found in the ship, all of different types, from rapiers to short, broad-bladed weapons. None were very fancy, and the find locations show that they were personal possessions. The navy offered no training in swordplay, but it was still common in Sweden that men carried swords and were expected to know how to use them.

There were only a few military muskets found on board, but the two companies of soldiers allotted to the ship had not yet been embarked. The muskets are of the type issued to the army in the late 1620s, before the reforms of 1631, with a heavy barrel and simple lock mechanism. Some members of the crew, probably officers, brought personal firearms with them. These are typical hunting weapons of German or Dutch style, smaller than the muskets and of finer manufacture. They may have been rifled, with spiral grooves down the barrel to make the bullet spin, increasing accuracy. Did the officers bring these in hopes of some sport, or with the intention of using them in battle? They might make good sharpshooter's arms, but the loading process was very slow.

THE MACHINE OF WAR

The political situation heated up in 1628. Imperial troops besieged the German Baltic port of Stralsund but could not close the sea approaches. Gustav Adolf agreed to aid the city and sent 600 troops by ship. The siege was lifted and Sweden had a stake in the main theatre of the great European conflict.

OLD ENEMIES AND NEW ALLIES

The normal tasks of the warships in the Swedish navy during the Polish war were to convoy troops and supplies to the front in Poland and to blockade the Polish ports, primarily Gdańsk. Swedish forces had taken the nearby port of Pillau (now Baltijsk, in Russia), and part of the war strategy was to divert Gdańsk's trade through Swedish hands in order to tax it. Smaller ships were used to cruise the approaches to Gdańsk and shepherd merchant vessels to Pillau. The blockade thus served not only to strangle Poland but to enrich Sweden.

1628 was an unusual year for naval strategy, as it was virtually the only year in the entire century that Sweden and Denmark were officially allied. Imperial troops had invaded Denmark in 1627 and threatened to attack the capital in the spring of the next year. That winter, Gustav Adolf and King Christian IV of Denmark met and agreed an alliance, in which the king of Sweden promised to send a squadron of warships to protect Copenhagen and the king of Denmark promised to use his control of the Öresund toll to divert trade bound for the Polish ports to Sweden.

When the campaigning season opened, the navy was divided into three main squadrons, plus smaller ships to cruise the Polish coast and carry messages between Stockholm and the different Swedish armies. The largest squadron would blockade the Polish coast, while eight ships would sail for Danish waters. A third squadron, of which *Vasa* was to be the flagship, would remain in Swedish waters as a reserve, awaiting developments on the front lines. As the year progressed, the situation changed dramatically. The emperor's troops left Denmark and marched on the Protestant towns of the German coast instead. They laid siege to the port of Stralsund, which called for aid.

Gustav Adolf redeployed his ships in the summer. The squadron assigned to Denmark was sent to relieve Stralsund instead, under Vice Admiral Klas Fleming. The forces on the Polish coast were increased, under Vice Admiral Henrik Fleming (Klas Fleming's cousin), and the reserve squadron was reduced in size, now that the likely need for ships was better understood. By the time *Vasa* sailed in August, the reserve or home squadron was down to four ships: *Vasa*, *Äpplet* (*Vasa*'s sister, laid down in 1627 but not yet finished), *Kronan* and *Gamla Svärdet*, supplemented by two small galleys. Erik Jönsson, having succeeded, more or less, in getting the new ship armed, was appointed Vice Admiral. He made *Vasa* his flagship and prepared to lead the squadron out to the summer fleet base at Älvsnabben, in the Stockholm archipelago. This would put the ships closer to the potential scene of any action, with a quicker exit from Swedish waters than if they stayed in the capitol. But events proved otherwise, and *Vasa* never left Stockholm.

Älvsnabben, in the southern archipelago, was the summer fleet base for the Swedish navy. It provided a protected harbour close to the main southern passage into the open Baltic. *Vasa* was bound here when she sank, and Gustav Adolf and his army left Sweden from here in 1630 to intervene in the Thirty Years War.

THE MACHINE OF WAR 65

The symbolic warship

1627

Mårten Redtmer had a story to tell, and not much time to tell it. While three hundred sawyers, carpenters and turners were building the king's new ship, Mårten and his handful of fellow sculptors had hundreds of carved and painted figures to make, as well as hundreds more smaller ornaments. They had started before the keel was even laid, and would be hard pressed to finish before the ship sailed. The cannon that Master Medardus was desperately casting and polishing were going to cost a lot more than Redtmer's carvings and would deliver a mighty punch if they were ever fired in anger, but there was a good chance they would never fire more than salutes. In all the years Redtmer had worked in Stockholm, there had only ever been one sea battle, and only two ships had managed to get a shot off. On the other hand, Mårten Redtmer's story would be heard any time the ship visited a port. People would see the guns, certainly, and they would speak of power, but the carvings would tell people who the king was, and what Sweden's new place in the world should be.

There had been a great deal of discussion of what sort of sculptures this ship should carry. It was a much bigger ship than the last one, *Tre Kronor*, launched in 1625, and it was to be much more ornate. The Crown was investing heavily in all aspects of this ship – new design, new armament – so a new level of decoration was no surprise. The ship was named for the royal family's own heraldic symbol, and so an important part of the story would have to be the king himself and his virtues. The current series of campaigns against the king's cousin, King Sigismund of Poland, was in its sixth year and at least partly based on their competing claims to the Swedish throne, so the ship should be a powerful propaganda weapon emphasizing Gustav Adolf's right to the crown. The Catholic Poles would also come in for some ridicule, which was always entertaining for the carvers.

The discussion had focused on the important ideas that would form parts of the composition, rather than the designs of individual figures. A few specific items were clear from the beginning. The figurehead should be a lion, as it would

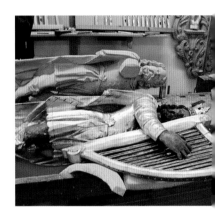

The Biblical King David and his harp are seen both on *Vasa* and the organ front in the German Church in Stockholm. The copy was executed using much the same technique as the original.

The carving work on *Vasa* was led by Mårten Redtmer, a German carver working in a typical north German style of the late Renaissance. He was employed on many projects in addition to ships, such as the magnificent organ front in the German Church in Stockholm in 1608. The current organ is a copy made in 2004; the original was sold in 1781 and can now be seen in two parts, in Övertorneå church and its neighbouring congregation at Hedenäset.

There was a common, sophisticated symbolic language in medieval and renaissance Europe, which was spread partly through pattern books. One such handbook of images which was circulating in Stockholm in the early 17th century was the Italian Cesare Ripa's *Iconologia* (1593).

be on any large warship, and the transom would be dominated by a large, gilded *vase*, the heraldic symbol of the king's family, for which the ship was named. There were also some thematic trends that had been popular in recent ships, such as references to the Roman Empire and the Goths from whom the king traced his decent, which would surely appear. The king himself identified with the Old Testament story of Gideon and his triumph with a small force over a more numerous enemy, so this element was certain to be included as well.

It was up to Redtmer and his fellow carvers to sort out the details. They had abundant experience of this type of carving. They had completed similar commissions on other ships, as well as churches and palaces, and the language of ideas had a well-established vocabulary in sculpture, a vocabulary transmitted not only in training but through pattern books as well. A figure of a warrior was not just a man in armour. How he stood, what he wore, what weapons he carried and how he held them meant something specific. Such images were a useful teaching tool for largely illiterate populations all over Europe, and those with classical educations would be able to read a great deal in the carvings. Complex ideas could be communicated if the composition was sound and the sculptors talented.

Once they could put tool to wood, they had to carve at breakneck pace, each stroke of the chisel or gouge taking off a big chip or shaving. No tap-tap-tapping with a tiny mallet, no slow refinement of the details. The tools flowed with power and confidence. Get the measurements off the ship – how tall should the figure be? How would it be attached? See the figure in the log and take away everything else. Capture the basic form and the pose with the axe and saw, hacking out chunks of wood. Add pieces for outstretched arms or where the log was not quite large enough. Rough in the plumes, the armour, the flowers. Set the edge just *here* and then *push* and *roll* just so to get the curve of the inside of a leaf, or the deep-set eye and steely gaze of a warrior. When it was done, admire it for a minute and then send it to be painted. It was always this way, whether they were working on a church, a palace or a ship. These big projects were always behind schedule and the money was always late.

GRAND IDEAS AND COARSE JOKES

Most of the sculpted figures were part of a panegyric, extolling the king's virtues and placing him in honoured company. At the bow, the long beakhead carried

ten figures of Roman emperors, about half life-size, on each side. These began with Tiberius, who was actually the second emperor. The first and greatest of the emperors, Augustus, was missing, but not by accident. His place was taken by the figurehead, a magnificent gilded lion over three metres long, carrying the Vasa fascine (the ship's name) in his paws. The lion was a traditional symbol of royalty and a specific emblem of the kings of Sweden since the Middle Ages. In addition, the Latin form of the king's name, Gustavus, was an anagram of Augustus. The message was clear: Gustav Adolf was the new Augustus, heir to ancient glory and the founder of a new empire. Roman imagery was used throughout the ship, with warriors in Roman armour appearing in several places. A revival of interest in the greatness of Rome had been one of the key features of the Italian Renaissance in the 14th and 15th centuries, and the fashion had spread northward, eventually far beyond the original Roman frontier. By association with ancient Rome, Gustav Adolf could be portrayed not as the son of a usurper, but as the upholder of the legal and philosophical traditions of the past. This did not imply any loyalty to

This coarse caricature of a crouching Polish nobleman, with distinctive tunic and hairstyle, could be appreciated by the crew when visiting the ship's heads.

THE SYMBOLIC WARSHIP 69

Earshell ornament is typical of the late Renaissance style popular in Sweden well into the 17th century, and can be seen in the decoration on houses in the Old Town of Stockholm.

The image is coarsely but powerfully carved, and typical of the other carvings on *Vasa*. About half of the carving was obscured by the bench on which the steersman stood.

The steerage is richly decorated compared to much of *Vasa*'s interior. Most striking is this monstrous face surrounding the ship's steering device, the whipstaff. The motif has roots in antiquity, such as the well known "Mouth of Truth" (*Boca della verità*) in Rome, which probably represents a river god.

The wide mouth accommodates the whipstaff even at large rudder angles.

Among the stern's rows of noble figures from history and myth are these grotesque faces, a common element in late Renaissance carving.

the modern Rome and the Roman Catholic church, in fact quite the opposite. The king, rather than the pope, was portrayed as the heir to Rome's greatness.

In addition to the figurehead and the emperors on the beakhead, the bow was decorated with the largest human figures found on the ship. Each side of the bow had a warrior figure over two metres tall, with a shield and drawn sword held behind his head. His armour, dress and hair are a mixture of Roman and Germanic features, possibly recalling the Gothic ancestors from whom the Swedes of the 17th century believed they were descended as well as the just cause for which Gustav Adolf claimed to fight. The pose of the warrior, with the sword behind his head and shield raised, is not aggressive but defensive and symbolized military might used to protect those not able to defend themselves.

The stern was even more elaborately decorated than the bow, and carried the most significant sculptures. It provided the largest canvas for the most complex compositions, including those that gave the ship's name and ownership. It was not yet common for a ship's name to appear as text, but it was usually represented by an image. The word "Vasa" does not appear anywhere on the ship or on any of the tens of thousands of objects found with it, but there is no doubt of the ship's identity.

The transom is a series of compositions in three main tiers. The lowest tier is the ship's name, in the form of a gilded *vase* on top of a diagonally striped shield. This is the coat of arms of the royal family. Although the *vase* had originally been a fascine, a bundle of sticks tied together, by Gustav Adolf's time it looked more like a sheaf of wheat. The shield is carried by a pair of cherubim and framed by cascades of fruit, a traditional symbol of prosperity. The cherubs carry olive branches, a symbol of peace, but are flanked by six life-size knights in full armour, holding spears and shields in regal but non-aggressive poses. The tier above is the royal arms of the kingdom of Sweden, a shield with the three gold crowns on a blue field quartered with a gold lion rampant on a striped field. The lion became part of the national arms in the 13th century through the Folkunga dynasty (it was their family symbol), but the original meaning of the three crowns has been lost. It has been the heraldic symbol of the kingdom of Sweden since the Middle Ages and is still used as an emblem on state property, such as embassy buildings and the air force's fighter jets. The Vasa arms are superimposed on the centre of the national arms, and the main device is carried by two gilded lions wearing crowns. A crown also appears above the central shield. These arms are reproduced in miniature on all of the ship's 24-pounder cannon.

The two lower tiers establish the ship's name and nationality, but the uppermost composition refers directly to the reason *Vasa* was built. A young man stands with his arms outstretched over eleven small figures, while two griffons hold a royal crown over his head. The young man is none other than Gustav Adolf himself. In 1604, Duke Karl of Södermanland was proclaimed king, making the deposition of his nephew Sigismund in 1599 legal and permanent. The griffon was the heraldic animal associated with Södermanland and became one of the heraldic symbols used by Karl. His son, Gustav Adolf, was invested as the crown prince, which thus excluded not only Sigismund but his descendants from the Swedish throne. Gustav Adolf thus received the legitimate right to rule Sweden from his father. The sculpture at the top of the transom, highest above the water, was a message to King Sigismund of Poland, Gustav Adolf's cousin: I am the rightful king of Sweden and you never will be.

The secondary figures around the main compositions of the transom continue the message of association begun at the bow. The Romans may have been great, but they had been conquered by the king's Gothic ancestors, who were thus even

The Swedish flag in its current form is of the same age as *Vasa*. Earlier it was alternating blue and white stripes. The cross flag was a Christian symbol referring to the cross and the triumph of the resurrection and became popular during the Crusades. The Swedish cross flag was initially used only on ships carrying the king or his representative, as here at the conquest of Riga in 1621.

THE SYMBOLIC WARSHIP 71

The figurehead, a gilded lion, holds the Vasa heraldic symbol in his paws.

greater. Gothic warriors guard the transom, together with the military heroes of the Old Testament. David is at the centre with his harp, just below the gilded *vase*, and Gideon's soldiers are also present. Classical mythology is not left out, as the corners of the transom are carried on the shoulders of two figures of Hercules, wearing the skin of the Nemean lion.

In addition to figures with specific meaning related to Gustav Adolf, there is a wealth of smaller carvings that were popular, generic images of the time. Grotesque masks and fanciful monsters abound, and these are part of a Germanic tradition going back into the Iron Age. The largest of these monster faces is inside the ship, in the steerage, where it frames the opening for the ship's steering gear. The whipstaff, which the helmsman used instead of a wheel, rises out of the monster's leering mouth as a four-metre-long tongue.

Considering that one of the official reasons Gustav Adolf had gone to war in Poland was to preserve Sweden's religious freedom, one may wonder that there are few overtly religious references. The Old Testament figures at the stern are the only biblical characters, and there is no clear sculptural depiction of the key tenet of Lutheran Protestantism, salvation by grace. There are sculptures which may refer to the Christian victory over death and in a moralizing way to Christian kingly virtues, but classical images are more common and the bulk of the sculptural program is humanistic rather than ecclesiastical. The Christian character of the king and his war may have been assumed and needed no explanation. It may be no coincidence that Gustav Adolf, who was himself a devout Lutheran, did not chose religious names for the ships in his navy, as other sovereigns of the period did. His ships were named after classical figures, animals, astronomical objects, or after the symbols of royal power.

Not all of the images were entirely serious or high-minded. Some were caricatures and jokes aimed at Sigismund's Polish supporters. Two of these were intended for the crew of the ship more than outside observers. At the bow, the two catheads (heavy cranes for lifting the anchors) rest on blocks carved in the form of Polish noblemen crouching under tables. To go under the table was to be humiliated, and the humiliation was magnified by the location. Not only was the Pole being crushed under the weight of the cathead, but the only place where a crewman could see the sculpture clearly was when he was sitting on one of the ship's two toilets.

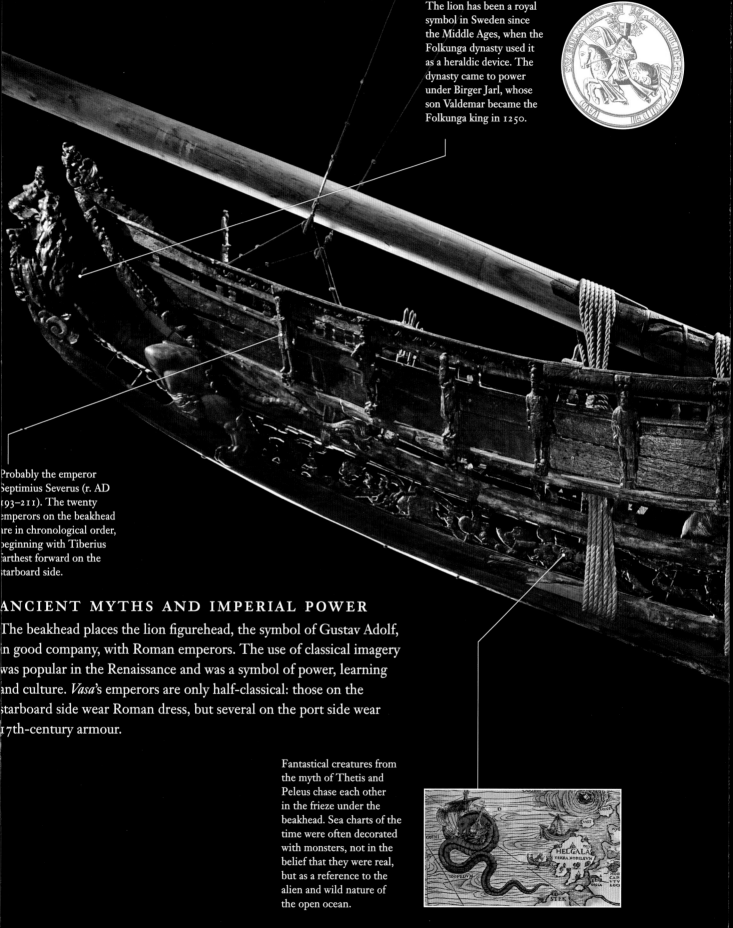

The lion has been a royal symbol in Sweden since the Middle Ages, when the Folkunga dynasty used it as a heraldic device. The dynasty came to power under Birger Jarl, whose son Valdemar became the Folkunga king in 1250.

Probably the emperor Septimius Severus (r. AD 193–211). The twenty emperors on the beakhead are in chronological order, beginning with Tiberius farthest forward on the starboard side.

ANCIENT MYTHS AND IMPERIAL POWER

The beakhead places the lion figurehead, the symbol of Gustav Adolf, in good company, with Roman emperors. The use of classical imagery was popular in the Renaissance and was a symbol of power, learning and culture. *Vasa*'s emperors are only half-classical: those on the starboard side wear Roman dress, but several on the port side wear 17th-century armour.

Fantastical creatures from the myth of Thetis and Peleus chase each other in the frieze under the beakhead. Sea charts of the time were often decorated with monsters, not in the belief that they were real, but as a reference to the alien and wild nature of the open ocean.

Tritons (mermen) were a popular maritime motif in ancient times. In myth, they were the sons of the sea god Poseidon and steered the waves.

THE LANGUAGE OF POWER AND POLITICS

Vasa's stern is actually the ship's "face," the most distinctive part of the ship and the part most often depicted in illustrations. It offered the largest canvas on which to tell the most important story about the ship and its owner.

The greater national coat of arms of Sweden looks much the same today. It represents the nation and its sovereign.

GARS, the initials of the king's title, *Gustavus Adolphus Rex Sueciae* (Gustav Adolf King of Sweden).

Six men and five women in common dress, the people for whom the king was responsible.

The stern lantern was placed here on an iron bracket, but was destroyed or lost in the sinking.

The main flag was hoisted on a staff mounted to the taffrail.

The griffon was a mythical beast, half lion and half eagle. They show that Gustav Adolf inherited the throne from his father, Karl IX, who was formerly the Duke of Södermanland and used the griffon as an heraldic device. The griffons crown a young, beardless Gustav Adolf.

The tent or drapery was an old symbol of princely status, from the tent erected over a throne.

This row of grotesques makes a startling break in the main sculptural program, but such faces were a popular late Renaissance image.

Gustav Adolf identified himself with the Old Testament story of Gideon, whose outnumbered force stole into the camp of their

VASA'S COLOURS

North German woodcarving of the late Renaissance was marked by bright, almost vulgar colours. Elites and the church used colour to mark their status and glory for the enjoyment of all.

Gilding was used on the most important elements, such as the lions holding the national coat of arms.

Greens and blues stood out from the general red/orange/yellow/brown background and were used to mark contrasts.

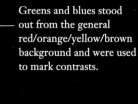

Bright yellow, from orpiment (an arsenic ore) was used instead of gilding in many places.

In this period, red was a royal or imperial colour, a sign of status. When Swedes later began to paint their wooden houses red, it was in a more prosaic attempt to copy the red brick of Dutch architecture.

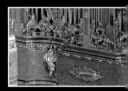

A trace of ancient designs can be seen in the egg-and-dart mouldings that decorate much of Redtmers work, both here and on his organ fronts.

In the human figures, an effort was made to use realistic colours as a base, with important deails picked out with gilding.

The original colours are still preserved in a few places, such as inside the mouth of this gunport lion.

The unpainted parts of the hull were tarred, which gave a reddish-brown colour, darkening over time towards black.

BRIGHT COLOURS

Carving was only the beginning of the process. Each figure and ornament had to be painted in order to have maximum effect. The sculptors themselves were probably the painters; in 1627, some of the shipyard painters complained that one of the carvers was meddling in what should have been painters' work. In the later Baroque period, shipcarvings were usually gilded and mounted on a solid-colour background, and some early reconstructions of *Vasa* portrayed her this way. Paint survives on many of the sculptures, and analysis of the remains has shown the reality to be far more lively. Human figures and foliage were painted in bright, naturalistic colours, with even more fantastic combinations used on mythical beasts and monsters. This polychrome style would have been old-fashioned farther south in Europe, more typical of the Middle Ages and Renaissance, and gave the ship a wild, festive appearance rather than the sober opulence coming into vogue elsewhere.

Most of the ship, from the gunports downward, was tarred and thus initially a deep reddish-brown, which would have darkened towards black if *Vasa* had stayed in service. The transom, the beakhead and the upper parts of the ship's sides were painted a vibrant red and served as the background for the sculptural compositions. The roofs of the quarter galleries at the stern were also red, with a fish-scale pattern mimicking roof tiles outlined in pale yellow. Pale yellow was also used for the railings at the upper edge of the sides, with the upper rails even lighter, almost cream.

On the carvings, reds, yellows and browns predominated, along with white, all pigments which could be obtained locally and cheaply. Much of the clothing on non-royal figures was in these earth tones. Yellows were also used to simulate gilding, often in the form of an ochre background with bright yellow and gilt highlights. Skin tones were in warm pink for men and very pale pink, almost white, for women. Blue was used on garments on royal figures, such as the Roman emperors, and some of the mythical creatures, such as the tritons and mermaids, used lavender and purple. Foliage and flowers were in naturalistic colours, so a fair amount of green appeared. The three-dimensional effect of the carving was emphasized by shadows and highlights applied to the base colour.

Although not as common as it would be later, gilding (covering the surface with thin sheets of gold) was still used for emphasis. The figurehead lion and the

The reconstructed colours on *Vasa* are fresh and almost naturalistic, but the motives are not from the local environment. They are the fruit, flowers and creatures of the Bible and classical mythology.

BEAKHEAD COLOURS

The ship's movement forward is emphasized by the sculptures of the beakhead. Bright colours and gilding make the figures stand out from the background and draw the eye to important ideas.

Tritons are painted in blues, greens and purples, sea tones recalling their fishy origins.

The figurehead lion, the largest sculpture on the ship, stands out, the sun glinting off his gilded hide.

The Roman emperors recall how ancient sculpture actually appeared, painted in realistic colours. These emperors make much use of blue, an expensive pigment associated with royalty and power.

Sharp contrast and bright colours attract attention to the bow and stern, while the sides are dominated by bronze cannon and snarling lion faces in naturalistic colours.

Warriors and their equipment are in realistic colours. The large warrior on the port side of the bow is missing, but surviving parts show he was a mirror image of the warrior to port.

two lions supporting the royal coat of arms of Sweden were almost entirely gilt, as was the large fascine in the ship's name panel. Gilding was added as emphasis on crowns, jewellery and other metalwork, and was used for the hair (including beards and moustaches) on the larger figures, especially Roman emperors and warriors.

The pigments used were primarily ground minerals, probably mixed with linseed oil as the carrier. Sweden's rich metal deposits provided a wide range of earth tones from various ores and oxides of iron and copper. White was commonly made from lead oxide, which was very stable and long-lasting, but poisonous. Black was one of the few pigments commonly made from organic materials, the soot or lampblack produced by burning. The best quality lampblack was produced by burning bone or ivory. Green could be made from copper compounds, but the most vivid green came from malachite, a mineral most common in Russia. Blue was the most expensive pigment, since few naturally occurring compounds produce a deep, rich blue. Smalt, a cobalt-rich copper ore, provided a locally available blue. The most vivid blue was made from ground lapis lazuli, a rare mineral occurring in significant deposits only in Russia and Afghanistan. None of these pigments were cheap or widely available and their use in substantial quantities on *Vasa* is a mark of the Crown's investment in the decorative scheme.

ARTISTS OR CRAFTSMEN?

Mårten Redtmer and his assistants were an unusual group among the workforce of the navy yard. They were not permanent employees hired by the Hybertsson brothers like the carpenters, sawyers and smiths building the ship. They were employed directly by the Crown and were not exclusively engaged in decorating ships. In later centuries, when the elaborate sculptural style of the Renaissance and Baroque had given way to a more restrained and simple style, shipcarvers were often maritime specialists making a living on the particular carvings needed for sailing ships, such as figureheads, nameboards and the simpler ornaments used at the stern. There were exceptions – the Swedish navy employed well-known artists in the late 18th century to carve the figureheads for its capital ships.

In the 17th century, men like Redtmer earned a living wherever carving was needed. This was possible because the style of ornament desired by the navy was not unique to ships, even if ship carvings probably had more mermaids, tritons

Roman emperors were common currency in European culture, and many claimed to be the "true" heirs of their legacy. The Renaissance spread knowledge of ancient art and learning, and *Vasa*'s carvers have gotten much of the detail right: the mantle, armour, and facial hair are all recognizably Roman in style.

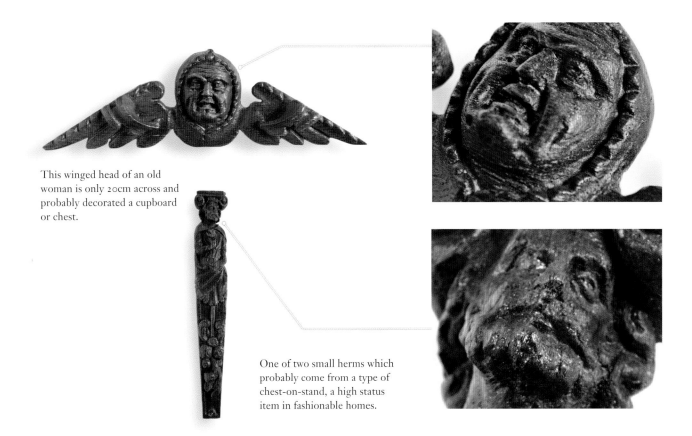

This winged head of an old woman is only 20cm across and probably decorated a cupboard or chest.

One of two small herms which probably come from a type of chest-on-stand, a high status item in fashionable homes.

There are also many smaller sculptures on *Vasa*, from the interior paneling and from fine furniture in the great cabin.

and seahorses. The same imagery and themes were used in palaces and churches, so carvers moved from project to project as patrons required. A ship was, in a sense, just another type of building, with difficult contours.

In Redtmer's case, some of his other commissions survive, such as the organ fronts in the German church in Stockholm, originals now in Övertorneå/Hedenäset, and Bälinge church in Uppland and the master for the bronze statue called Kopparmatte in the Municipal Court Building in Stockholm. From these it is possible to recognize his work on *Vasa*. He seems to have had a limited repertoire in human figures, carving the same face over and over: deep-set eyes, prominent cheekbones, and a beak of a nose, usually complemented by a curly beard and moustache. The general style of his carving is typical of northern Germany, where he was probably trained, and he can be associated with many of the best quality carvings on the ship. A few other carvers working at the shipyard are known by name, from their appearance on payroll records, but cannot be associated directly with specific carvings. After Redtmer, the most senior were probably Hans Clausink, another German from Westphalia, and Johan Thesson, a Dutchman. Redtmer and Clausink were both listed as masters in the 1620s. Two other carvers, Giärdt and Petter, are known only by their first names. There must also have been assistants in order to complete such a large project on time.

Each carver had his own history and style, even though they all worked in a similar vocabulary of stock images. The northern German style of Redtmer is the most common, but at least one carver worked in a distinctly different Flemish-Dutch style (probably Thesson). Redtmer's work is very modern for northern Europe in the 1620s, and he was one of the leading exponents of the lush, new style of carving that became popular in Stockholm in the 1630s and 1640s. Other carvings are relatively old-fashioned for the 1620s, in a heavier and less accomplished technique. This has been interpreted to mean that the carvings themselves were old, carved years before the ship was built and stockpiled, but it may be simply that some of the carvers were older men, working in the style in which they had been trained decades earlier.

Because the carvers, along with the joiners who made the interior panelling and furniture for the cabins in the stern, were paid separately from the rest of the navy yard personnel, they appear as a separate item in the building account compiled after the ship sank. The total amount paid for carving and joinery was 4,600 dalers, 8.6 percent of the total hull cost of 53,300 dalers and nearly as much as all of the Swedish carpenters working on the ship earned. It was 20 per cent of the labour total. A similar building account survives for *Tre Kronor*, the smaller ship launched in late 1625. The overall distribution of costs is approximately the same for the two ships, except for the expense of carving and joinery. For *Tre Kronor*, it was only 6 per cent of the total cost of 21,350 dalers and 13 percent of the total labour cost, which suggests that *Vasa* was more richly decorated. The ship was intended not only to set a new standard for firepower, a physical warship capable of outfighting any other ship in the Baltic, but it was also intended to be an unmatched metaphysical warship, a floating symbol of Gustav Adolf's ambitions for himself and Sweden.

Mårten Redtmer's work is often easy to recognize in his carved faces, which are consistent in their features. He worked swiftly and confidently, and created powerful images with the fewest possible strokes of the gouge.

The full-rigged ship

1628

The carpenters in the shipyard had made the masts and tops and the turners' shop delivered all of the blocks, deadeyes, parrels and other tackle on time, one of the few things not behind schedule on this ship. There had been some delay in getting all the cordage in, and the material that was finally delivered was of very mixed quality. The Clercks normally provided rigging to the fleet and could usually be relied on for timely delivery and decent quality, but they had not supplied the rigging for *Vasa*. Some of the rope was excellent, some was only workmanlike, and some was barely usable. In the worst, the hemp had not been properly hackled, so it was full of hurds and dirt, and it was very uneven in diameter and tension, so it tried to tie itself in knots. Lieutenant Petter Gierdsson had the task of trying to put all of this hemp, wood and iron together into a functioning rig. He was not worried about getting the ship ready in time, since he had his responsibilities well in hand, but he may have been the only one in the navy yard not feeling the pressure as it ratcheted ever upward that spring. He was a little concerned that some of the rope was not going to last very long and that he might be rerigging the ship in a year's time.

Normally, it would be Captain Söfring Hansson's responsibility to rig the ship, since he was the *gårdskapten*, the naval officer assigned to supervise the work in the Stockholm navy yard. He had not had time for this sort of practical task since Master Henrik Hybertsson died over a year ago, in the spring of 1627, and his real ability lay in administration in any case. He had gradually assumed greater control over the yard. Henrik's brother, Arendt, had tried to help run the yard, but he was always travelling. The situation was unworkable, and so the navy had stepped in and put Captain Söfring officially in charge of the navy yard back in the winter. He was already too busy in the summer of 1627, so he had delegated the job of rigging *Vasa* to his lieutenant, who had not let him down.

Lieutenant Petter Gierdsson had worked with the ship's master, Jöran Matsson, to coordinate the men, materials and tools needed for the complicated and

Hemp was a strategic material in 17th-century Europe, and Sweden controlled one of the primary sources for it after 1621. A ship the size of *Vasa* required tons of it for rope and sails. Here, the reinforcing boltrope from a sail is seized at the corner with lighter line to form a loop.

The reconstructed *Kalmar Nyckel* of Wilmington, Delaware sails with a similar rig and steering system to *Vasa* and can offer useful clues to how *Vasa* was sailed. The original ship was built in Holland in 1625 and in 1638 carried the first Swedish colonists to North America.

tedious task of setting up the ship's enormous masts, yards, sails and rigging. Because of the size of the objects to be moved, each stage had to be planned and managed carefully. The largest component, the lower mainmast, weighed over 15,000 pounds (6.5 tons) and was nearly 90 feet (26 m) long. Even relatively small objects, such as the upper yards, had to be swayed up carefully, without fouling any of the rigging already in place. Many men were needed for hauling lines, heaving round the capstans and guiding heavy objects past obstructions, but relatively few of them had much experience with this kind of work. Matsson's knowledge had been valuable, but he had spent much of his time in the spring of 1628 in the hold, supervising the loading and placement of the stone ballast that would allow the ship to carry her sails. Petter was left to direct the work on deck and aloft.

What did worry the lieutenant was the ship's tenderness. He could not help noticing that the ship rolled too easily, which had not made rigging any easier. He had to be very careful how he stowed spars and rope on deck in preparation for rigging and how he managed the weights of the heavier items when swaying them aloft. He knew Matsson was busy down in the hold, trying to get enough ballast in to keep the ship upright when the wind filled her sails, but as the ballast and guns came on board and the ship sat ever deeper in the water, she did not seem to get any stiffer. They would not rig the topgallantmasts and sails, the uppermost parts of the rig only used in light winds, since these were not normally needed in local waters. They would only bend on the handful of sails needed for normal manoeuvring in the confined waters of the Stockholm Archipelago. Old Captain Hans Jonsson, who was scheduled to command *Vasa*, had plenty of sailing experience and would surely know how to handle a tender ship.

HEMP AND CANVAS, MUSCLE AND WIND

Vasa's hull and armament may have been innovative, but the rig Petter Gierdsson assembled was entirely conventional, consisting of ten sails on four masts and the bowsprit. The division of sail area meant that no one sail was too large to handle with human muscle power and different sail combinations could be used to suit different wind directions and strength. Composed almost entirely of square sails, bent onto horizontal yards, the rig was potentially very powerful but not especially weatherly. This meant that the ship could sail across the wind or before

Vasa's mainmast (the lower mast; the upper parts do not survive) was made of 11 pieces of wood plus reinforcing iron bars and straps. It was 26.5m long and weighed over 6.5 tons.

it, but not as far into the wind as ships of the 19th century could. This limited a captain's options, and routes had to be planned to use the available wind or the ship had to wait until the wind changed.

Most of the rig's power came from the foremast, at the bow, and the mainmast, in the middle of the ship. Each of these masts was in three overlapping sections to achieve its full length: a lower mast, topmast and topgallantmast. Each mast carried its own sail: course, topsail and topgallantsail. The topsails were the main driving sails, together with the fore course. The main course could be set for more power, but was not used as often, since it blanketed the fore course on many points of sail and obstructed the view of the deck. It was still important in some situations, such as in storms, but it was usually furled in battle, and in *Vasa*'s case, was never bent on to the yard. It was found in the sail room below decks, still neatly folded and tied up in a bundle together with all of the lighter sails, when the ship was excavated in 1961. The topgallantsails were only set in light winds, and the topgallantmasts might be stowed on deck or lashed to the ship's side until the sails themselves were needed.

Vasa had only four sails bent onto the yards when she sailed. The rest, together with the mizzen bonnet and two sails for the ship's longboat, were stored in the sail room on the orlop, where they were found in a sodden, muddy heap when the ship was excavated in 1961.

THE FULL-RIGGED SHIP   85

The ten-sail ship rig was typical of large vessels in northern Europe between about 1570 and 1670, but not all sails were used equally. Most normal sailing was accomplished with just four: the fore and main topsails, the fore course and the mizzen. The main course could be added for power or the spritsail for balance.

The ship's smallest sail, the fore topgallantsail, is the best preserved and is exhibited in its entirety.

One normally set sails in a particular order, depending on how the ship got under way. The circle shows the location of the total centre of effort.

The fore course was normally set first, which would turn the ship away from an anchoring spot or a quay and allow the ship to sail directly before the wind.

The mizzen would be set to balance the helm and allow the ship to manoeuvre.

The topsails were added to provide the main driving power, but would cause the ship to heel (lean) more.

The spritsail topsail was typical of the 17th century, but was only useful in a few situations.

The spritsail could provide both power and helm balance on the right point of sail.

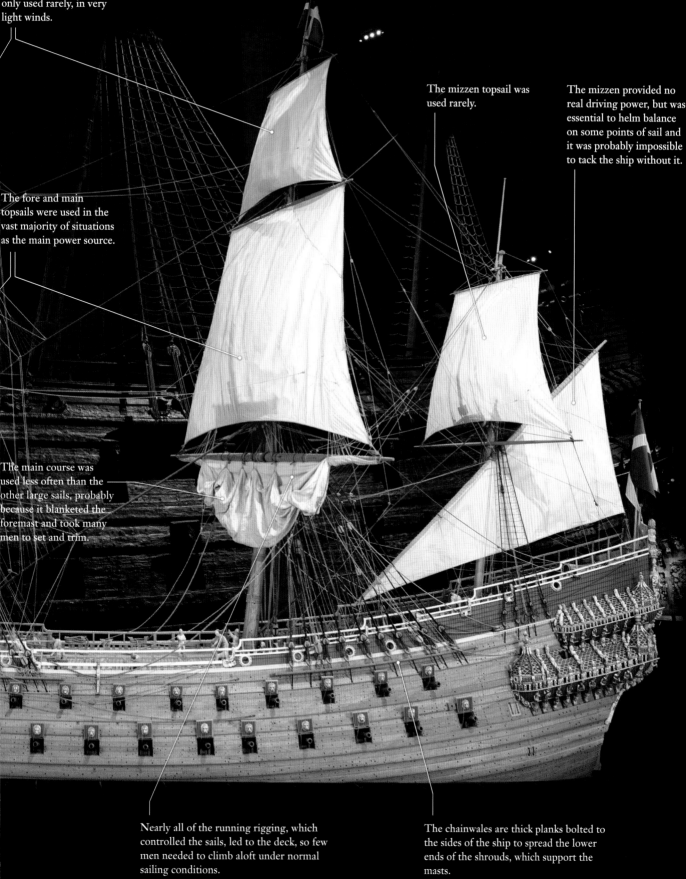

The tiller is just over 9m long and provided plenty of leverage to move the rudder, which weighed over 2.5 tons. *Vasa*'s rudder had a maximum deflection of 22 degrees to each side, which was more than enough for most manoeuvres, as long as the sails were properly trimmed.

The other two masts carried sails that did not, for the most part, contribute significantly to the ship's speed, but allowed the crew to balance the wind force applied to the hull so that it was not difficult to steer. On different headings relative to the wind direction, the centre of effort of the sails changes relative to the centre of resistance of the hull, increasing or decreasing the effort needed at the helm to maintain a steady course. By adding or removing sails at the ends of the ship, the crew could adjust the centre of effort to minimize the force needed at the helm. The mizzenmast, at the stern, carried a triangular sail set fore-and-aft, the mizzen, and a small square topsail on a topmast. The mizzen was frequently set to balance the helm, but the mizzen topsail seems to have been rarely used. At the bow, a long spar called the bowsprit projected forward. A relatively large

sail, the spritsail, was carried on a yard under the bowsprit, and it could provide significant power on the right point of sail or make a noticeable difference in helm balance. A smaller sail, the spritsail topsail, was carried on a small mast set on the end of the bowsprit, the spritsail topmast. It offered little power, was tricky to set effectively, and was less often used; it disappeared at the beginning of the 18th century, replaced by more effective fore-and-aft sails set on the stays between the foremast and bowsprit.

The sails were an important part of steering the ship. By shifting the centre of effort fore and aft, the ship could be made to balance at different angles to the wind. If the balance was not right, then the helmsman would have to keep the rudder well over to one side to correct for the imbalance. This was a lot of work for the helmsman, and it slowed the ship, since high rudder angles created unwanted drag through the water. In a large ship such as *Vasa*, the forces acting on the rudder were also potentially large, so the helmsman needed some mechanical advantage over the ship. In smaller vessels, this was provided by the tiller, a long lever attached to the head of the rudder. On *Vasa*, the tiller is an oak timber over nine metres long and weighing more than a third of a ton, located on the lower gundeck. A helmsman standing at the tiller would have no view of the outside world. The ship's wheel, the device used for moving the tiller on modern ships, had not yet been invented, and from the end of the Middle Ages until the early 18th century, the solution was a simpler device, the whipstaff. This made it possible to place the helmsman a deck higher than the tiller and allowed him to move the rudder over a wide range without having to walk back forth with the tiller. At small rudder angles, it also provided some mechanical advantage for shifting the weight of the rudder and tiller, although this disappeared as rudder angle increased.

The whipstaff was a long wooden rod with an eye at the lower end, which fitted over the end of the tiller. The shaft passed through a wooden bearing mounted in the upper gundeck, providing a fulcrum. As the helmsman pushed the whipstaff to one side, it pushed the tiller to the opposite side and thus the rudder to the same side as the whipstaff. To turn to port, the helmsman pushed the whipstaff to the left. As the tiller turned, its end moved farther away from the bearing, so the whipstaff had to slide in the bearing. In practice, the helmsman had to push the whipstaff down, through the bearing, at the same time that he

The steerage was on the deck above the tiller and the tiller was shifted with the whipstaff, pivoting in a bearing, the rowle, mounted in the deck. An iron-reinforced eye at the lower of the whipstaff connected it to the tiller.

THE FULL-RIGGED SHIP 89

The art of making rope by hand or with human-powered machinery is almost forgotten today. The Danish ropemaker and researcher Ole Magnus studied *Vasa*'s rope and could see that it was of highly variable quality.

leaned it to one side. The system provided adequate leverage to allow one man to manage the ship's heavy rudder, but it was hard work, moving back and forth, pushing and pulling on the whipstaff. The helmsman's view was still restricted. He could see a few sails, so would know if they were properly filling with wind, and he could see the compass mounted in a binnacle in front of him, so that he could maintain a steady course, but he could not see what was in front of the ship. He depended on others to warn him of hazards and the ship's officers to provide course orders.

CONTRACTS, DEALS AND DEBTS

The Hybertsson brothers were contracted to build *Vasa*, but not to rig her. They provided all of the wooden components of the rig, from the masts to the tops (the round platforms at the mastheads) to the blocks, deadeyes and other movable parts. They also were responsible for the working parts of the rig built into the ship, such as the carved knights at the masts. The surviving hardware shows that the shop producing these components was highly specialised, using standard patterns to turn out large numbers of identical components. The rope and sails for the navy were normally provided under a separate contract with another family of foreign entrepreneurs.

The Clerck family came to Sweden from Scotland early in the 17th century, originally to serve in the navy. Rickard Clerck rose to the rank of admiral, but saw good business opportunities in the procurement side of naval operations. When the *arrende* system of private contracts replaced direct Crown administration of the fleet's material needs, Clerck signed the first such contract, in 1615, for the maintenance of the rigging and sails. He renewed the contract in 1620, and after he died in late 1625 or early 1626, his kinsman Hans (sometimes written Johan) Clerck succeeded him. His descendants continued to serve as admirals and have a hand in the fleet's administration through most of the rest of the century.

In August of 1626, Hans Clerck agreed to maintain the ships of the fleet for a period of four years, at an annual cost of 2 dalers per last, a measurement of ship capacity (about three cubic metres) commonly used in Scandinavia and the Baltic. The contract listed the 30 major warships in the fleet by name and size. This is the first written mention of *Vasa* by name, at 400 lasts the largest ship in the fleet. The contract specified what Clerck was obligated to provide and the quality of

In *Vasa*'s time, rope was often made in the open air; covered or enclosed ropewalks became common later. When the Swedish navy's main base was moved to Karlskrona in the 1680s, the ropewalk was one of the first buildings built, and it remained in use until the 1960s. Over 300m long, it has been put back into limited operation as a museum exhibit in the 2010s.

materials he was to use. Rope and cable was to be of the best Riga or Königsberg hemp, and sails were to be sewn of French sailcloth. Anchor cable lengths were specified based on ship size, as were the number and type of flags and streamers. *Vasa* should have had four anchor cables 110–120 fathoms (196–214m) long, five flags and three streamers (long banners flown from the mastheads on special occasions). Clerck's responsibilities in the event of the sinking of a ship were spelled out, including if some of the rigging should be recovered in a usable state. These provisions should have applied after *Vasa* was lost, as her topsail rigging was probably salvaged soon after the sinking. The Hybertsson brothers' contract included new construction, but Clerck had to settle the terms of provision of rigging for new ships by separate contract. He was given control over the Crown's ropewalk in Stockholm (located near what is now Norrmalmstorg, on the mainland just north of the navy yard), and a specific sum of money to cover the salaries of the ropemakers there.

In the normal course of things, Clerck should have had the rope for *Vasa*'s rigging laid up in the ropewalk and the sails sewn in his sail loft from materials

THE FULL-RIGGED SHIP   91

SAILING *VASA*

Sails had to be set and trimmed in coordination with the helm, or it would be difficult or even impossible to steer the ship. When changing directions, sails had to be adjusted with careful timing or the manoeuvre might fail, leaving the ship dead in the water or drifting towards a lee shore. Changing tacks, either by tacking or wearing, called for especially careful cooperation and coordination.

The braces, attached to the ends of the yards, were used to turn the sails relative to the ship and the wind.

The upper edge of the sail was tied to the yard with robbands, spaced about 20cm apart.

The clews (lower corners) of the topsails were hauled out to the ends of the course yards by sheets.

Vasa probably sailed best with the wind coming from the quarter (from the stern and slightly off to one side); in the right conditions, she might have been relatively fast, due to her narrow beam and streamlined lower hull. She could probably sail into the wind, but would have made considerable leeway (sideways drift).

There were two ways for a ship to shift the wind from one side to the other:

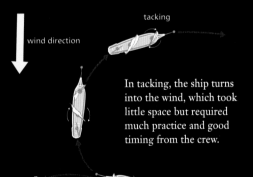

In tacking, the ship turns into the wind, which took little space but required much practice and good timing from the crew.

In wearing, there was less risk but it required more searoom and the ship lost ground. It was the best option in hard weather and the safest with an inexperienced crew.

The clews (lower corners) of the fore and main courses were controlled with sheets (leading aft) and tacks (leading forward).

The bonnet was an extra strip of sailcloth laced to the lower edge of a sail to provide more power. It could be removed in heavy weather.

he had imported from Livonia and Holland (the principal market for French sailcloth). He would deliver the finished goods to the shipyard, where he kept a warehouse for the storage of rigging. He was not responsible for rigging the ship – by long-standing custom, this was the task of the navy, the people who would sail the ship.

But as with so much else in the history of the construction of *Vasa*, things did not go according to plan. Two Dutch entrepreneurs running the shipyard in Västervik, Paridon von Horn and Christian Welshuisen, had built a large warship for the Crown under a contract originally signed in 1619 and amended several times. The ship, *Äpplet* (the Apple) was delivered in 1622 as a completed hull. The Crown paid Rickard Clerck 6,500 dalers for the rigging, and the ship eventually entered service. She was found to be unfit, unable to carry her guns. The king signed a contract with the Hybertsson brothers, at about the same time as their *arrende* in the winter of 1625, to rebuild or improve the ship, but in the end von Horn and Welshuisen were compelled to buy the ship back from the Crown. This meant repaying not only the 33,000 dalers they had received for the hull, but purchasing the rigging as well. They took the ship back, after her armament had been removed, and operated her as a merchant ship for a short while before selling her to a group of merchants in Holland.

Like all the other entrepreneurs with *arrende* contracts, they were chronically short of cash and chose to repay part of the debt in kind. To cover what they owed on the rigging, they offered to provide the rigging for the new, large ship under construction in Stockholm. This was deemed acceptable, and they delivered the rigging for *Vasa*. They did not have Clerck's connections or established facilities to call upon, and so they probably had to contract with a number of ropewalks and workshops to complete the order. They may have made a separate arrangement with Clerck concerning the details, since Clerck was obligated for the maintenance of what they provided, and he may have assisted with contacts or even labour.

The surviving rope is very uneven in quality, clearly the product of many different manufacturers rather than a single ropewalk. The raw material, hemp, is of good quality but not consistently processed. Hemp for naval use was imported into Sweden as raw fibre, which had only undergone the first stage of preparation, retting, to separate the fibre from the woody plant stalk. The remainder of the processing, which included scutching to remove the outer casing of the fibres

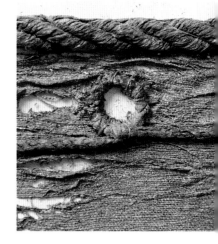

Detail of the head (upper edge) of the main course. The headrope, sewn to the doubled-over edge of the sail, reinforces the head, while the holes for the robbands tying the sail to the yard are formed by loops in a continuous line, sewn to the face of the sail. This technique was replaced in the later 17th century by individual rope grommets at each robband hole.

A downrigged ship (such as this English ship taken by the Dutch) was a common sight in the winter, when all but the lower masts was taken down for storage and maintenance. This is how *Vasa* is displayed today.

and hackling to comb all of the unwanted dirt and detritus (hurds) out of the fibre before it could be spun into yarn, was carried out in Sweden. Spinning and laying up of rope was carried out in semi-mechanised ropewalks, using machinery developed in the late Middle Ages, which allowed very long yarns and thus long ropes to be made efficiently. Several ropewalks were in operation in Sweden in the 1620s, in addition to Clerck's ropewalk in Stockholm, so von Horn and Welshuisen had a number of potential sources. Some of the rope is laid by hand, in an ancient technique, and probably represents short lengths made up by the crew while rigging the ship.

The sails, on the other hand, are of very high quality semi-finished materials and more consistent workmanship. The sailcloth is not French, but Dutch, the usual second choice for the fleet (French cloth was very expensive), and a weaver's tally markings are still visible on one of the cloths. The width of the cloth suggests it was woven in Krommenie, near Amsterdam, a major centre for the production of naval sailcloth. The boltropes sewn into the edges of the sails are of a special type made for sail roping, of consistently high quality, and may have been purchased abroad.

Von Horn and Welshuisen considered their debt paid, but the Crown did not. The admiralty valued the rigging for *Vasa* at only 6,000 dalers, and demanded the other 500 dalers in cash. The Västervik entrepreneurs refused. Letters went back and forth and the dispute dragged on into the 1630s without resolution, long after the ship and most of its rigging were lost.

POWER AND SKILL

The expansion and reorganization of the military under Gustav Adolf strained both the army's and the navy's traditional recruiting systems. To meet the growing need, a new system of conscription was instituted, taking one of every ten adult males from each district for military service (one of twenty in districts owned by nobles). By the later 1620s, when *Vasa* received her crew, over 85 per cent of the navy's seamen were conscripts. In districts designated for naval conscription, the officials responsible were charged with finding men with maritime experience, but this requirement probably could not be met consistently. The overall population was not very large, and the maritime sector too small to support the manning requirements of an expanded fleet without causing significant problems in fishing

There are few contemporary pictures of ships like *Vasa* being rigged. Here, the crew of the Polish ship *King David* have taken down the main yard and are busy bending (tying) on the mainsail.

and commerce. The population felt that conscription was a much heavier burden than the higher taxes that accompanied Gustav Adolf's wars, and was not willing to send its best men to the military. As a result, it was not possible to be sure that new seamen had much relevant sailing experience when they arrived in Stockholm for assignment to particular ships. There was no central training facility, so the new men had to learn their tasks on board. This presented both technical and organizational problems, which were solved by a combination of crew training and rig design.

The typical naval ship rig of the 1620s was, compared to the merchant rig of two centuries later, relatively complicated and labour intensive. There was little machinery to assist in moving heavy loads, and the weights involved were considerable – the maintopsail and its yard weighed over half a ton. The deck space available for handling the rigging was limited, due to the ship's narrow beam and the obstructions created by the upper deck guns. There was plenty of muscle power available, since there were supposed to be a total of 90 ordinary seamen, divided into two watches, in the ship's sailing crew of 133 but not necessarily much knowledge. To make this work, the rigging was optimized for a large crew of unskilled labour directed by a few experienced men. Most of the

Men walk round the capstan on the upper deck of a scale model of *Vasa*. This capstan was used for handling the rigging of the foremast and for catting the anchor. The other two capstans were used for handling other anchor lines and the rigging of the mainmast.

lines needed for ordinary sail setting, trimming and furling were led to the deck, and only a few men were needed to climb aloft. Many of the features that made work aloft safer and more practical, such as footropes slung under the yards to give crewmen a place to stand while hauling in handfuls of sailcloth, had not yet come into use. Sails were not furled by taking them up to the yards evenly, as was common a century later, but by bundling them in toward the mast. The topsails were furled into the tops, the large round platforms just below the topsail yards, so few men had to crawl out along the yards.

Down on deck, the cordage of the running rigging (the ropes handled in sailing the ship) was led so that many men could grab hold and pull in a clear deck space. Each watch had a seasoning of experienced men who knew which lines to haul and when, and it was their task to teach the new men and provide leadership, so that all would pull together. This type of rig put a premium on the management skill of leaders, in order to use brute labour effectively. The officers and non-commissioned officers had to know how to arrange the men needed for any particular operation and to make sure that orders were given in the right order to the right people. As economic forces on merchant shippers encouraged a steady reduction in number of men and increase in ship size, each member of the crew needed to be more knowledgeable. The role of human muscle was replaced by machinery, until sails themselves were replaced by steam. Naval vessels, because they had to carry large crews for military reasons, had no reason to economise on labour and so continued to sail ships with old-fashioned, labour-intensive rigs well into the 19th century. The ability to manage large numbers of unwilling men compelled to military service continued to be a prime requisite for naval officers.

Even with a large crew, some tasks were too big to be accomplished by unaided muscle. Weighing the 1.5-ton anchor and setting up topmasts, which could be housed (taken part way down) in bad weather, required the use of the capstans. These human-powered machines were mounted where lines could be taken directly to them. The main capstan, on the lower gundeck, was used for hauling in the anchor, sending up the main topmast, and raising the mainsail. The anchor cables were too large and stiff to be wound around the capstan, so they were tied to an endless loop of rope, the messenger, which could be wound several times around the capstan and passed around a pair of rollers at the bow to return down the other side of the deck. Smaller capstans on the upper deck and upper gundeck could handle the lighter running rigging of the foremast and secondary anchoring tasks. Bars placed through a capstan allowed up to 32 men to walk around it, applying concentrated force and leverage.

Under way, *Vasa* was potentially a fast ship, due to her relatively long, narrow hull and shallow draft. In the right conditions, she might have exceeded 10 knots, but some of the features that made her fast also made her tender. She was not rigged for heavy seas or bad storms. Her sails were lightly made, with little reinforcement at wear points, and she lacked extra stays and standing rigging to support the masts in a blow. English ships sailing the Atlantic Ocean were starting already in the 1620s to adopt these improvements, and Dutch riggers would soon begin to make the same changes.

Vasa's rig was perfectly adequate for the Baltic in the normal sailing season of May through October. Storms can build up quickly, but are not usually of the same ferocity as those encountered in the North Sea. As the 17th century wore on, strategic concerns forced the Swedish navy to sail later and later into the autumn and winter, until by the end of the century it was not uncommon to operate year round. Backstays, reef points and footropes would be essential.

The upper gundeck capstan on *Vasa*, as it was found. It has a rocking pawl pivoting on the deck in front of it. This could be turned into the capstan to prevent it from rotating in one direction or the other. Other capstans had a pair of pawls.

A floating community

1628

No muster list of the crew survives, if one was ever made. Navy planning documents for the 1628 campaigning season report how many should have been in the crew, but there is no way to know how many actually boarded the ship. A handful of officers who survived the sinking were called to testify at the inquest that followed, and so their names and something of their history is known. One of the dead is mentioned by name, and two ordinary seamen appear in the archives, but the crew is largely anonymous. What we know of them must come from their remains, their possessions, and the space in which they lived and worked.

As a group, the crew were a reflection of Swedish society in general. The ordinary sailors were mostly conscripts, and came from coastal districts and towns all over Sweden and Finland. They probably grew up in farming or fishing families, with a leavening of labourers and craftsmen. They likely counted a few foreigners as well, Germans or Dutchmen who had entered the international market for maritime labour or Danes and Norwegians who had changed sides, willingly or unwillingly, in the constant wars between the two Scandinavian powers. The officers were more likely to be ethnic Swedes with better economic backgrounds, men who sought the status and opportunity that service to the state offered under Gustav Adolf, but foreigners were still taken into Swedish service on occasion.

The divide between officers and men was reflected in their accommodations, food and personal possessions. The officers lived in the finely furnished cabins in the stern, where they sat on chairs and benches at a table to eat from pewter plates and drink from glass tumblers. They slept in beds with mattresses, and could look out at the world through glass windows, an expensive luxury in 17th-century Sweden. The men slept on the decks, between the cannon. They ate and drank there as well, from wooden bowls and tankards shared among them. Their view of the world outside, if they were not on deck working, was the fitful light that filtered down through the gratings in the decks above.

Gratings in the upper decks let light and air down into the gundecks and could be covered in rough weather.

A mess of six to eight men ate from a common bowl and shared a common sleeping space on the deck between the guns, but spoons were a personal possession, sometimes carved by the sailors themselves and decorated with paint or incised lines.

1628

THE GREAT CABIN

Fitted out and finished like a room in a royal palace, the great cabin provided accommodation for up to eight people on fold-out double beds built into the settees on either side. The high-ranking officers or guests who used the cabin had access to a private gallery in the stern and had the luxuries of a floor of wide pine boards, glass windows and movable furniture. In addition to a plain table and chairs, there were also richly decorated cupboards.

A wide-brimmed felt hat with silk band was high fashion in 1628 and popularized by the king. It was found in a chest with spare shoes and slippers, mittens, money, a little keg of spirits, and a sewing kit. It probably belonged to one of the higher-ranking petty officers.

The crew was divided into two groups, mariners and soldiers. The former sailed the ship and were on board for the duration of the sailing season. The soldiers defended the ship and were only embarked if the ship was deployed to an active combat theatre. Because *Vasa* was bound for the summer fleet base at Älvsnabben in the southern archipelago as part of the home or reserve squadron, the two companies of soldiers planned for the ship were not yet on board; they would probably wait on land until needed. A company for shipboard service had an official strength of 150 officers and men, although few companies managed to maintain this level in service due to recruiting difficulties, sickness and desertion. These 300 or so men made up just over two thirds of the total crew, and were expected to stay out of the way when the ship was under sail.

The manning lists for 1628 specify 133 mariners for *Vasa*. Three were commissioned officers, the captain and two lieutenants. Most ships only carried a single lieutenant, and only the three largest had two. One lieutenant, Petter Gierdsson, testified at the inquest, but nothing is known of the other.

Hans Jonsson, often called "Old Captain Hans Jonsson," was originally appointed captain on *Vasa*. He held the rank of senior captain (*överkapten*) and was one of the most experienced commanders in the fleet. He was replaced as captain before the ship sailed by Söfring Hansson, who had supervised the ship's construction. He was a Dane by birth, and was often called "Söfring Jute." He had entered Swedish service in 1602 and commanded several ships over the next 15 years or so. By the later 1610s he was the *gårdskapten*, the supervisor of the navy's interests at the navy yard in Stockholm, a post he held into the 1630s. He had few sea commands after 1620, probably because the navy valued his administrative abilities. The loss of *Vasa* does not seem to have hurt his career, as he was eventually promoted to *holmmajor*, a sort of rear admiral, and occasionally had sea commands in the 1630s. Old Hans Jonsson fared less well. He was on board when the ship sailed as a guest or consultant and was trapped below decks. He is the only casualty known by name, as the report to the king immediately after the loss specifically mentions him. This suggests that the king knew him personally.

The officers' commands were communicated to the crew through the non-commissioned officers. The ship's master (*överskeppare*), Jöran Matsson, was a professional mariner assigned to the ship in a more permanent fashion than the

Upstairs, downstairs. The plainly furnished upper cabin could be reached from the great cabin by a hidden staircase in the quarter gallery.

captain, and stayed with the ship over the winter, supervising maintenance and preparing for the next sailing season. Matsson was eventually commissioned in the 1630s and rose to the rank of captain, becoming something of a specialist in the raising of sunken ships. The master was assisted by a master's mate (*underskeppare*). A boatswain (*högbåtsman*) and a boatswain's mate were responsible for maintenance of the ship and provided leadership to the ordinary sailors. The boatswain, Per Bertilsson, survived the sinking to testify at the inquest. In battle, he directed damage control. Two pilots (*styrman*) were responsible for navigation. They were expected to be familiar with the waters in which the ship normally sailed and to know the winds and currents of the Baltic. They were important men, paid as much as the ship's master, the senior non-commissioned officer on board.

The sailing crew consisted of 90 ordinary seamen, divided into two watches. Each watch was supplemented with a leading seaman (*skeppman*) and two leading seaman's mates, experienced sailors who could teach new recruits and conscripts.

The helmsman was serenaded by a group of seven cherubs playing musical instruments, sculptures mounted on the bulkhead behind and above him in the steerage.

A watch was commanded by one of the lieutenants, with the master or master's mate and boatswain or boatswain's mate to assist and a pilot to navigate. For command purposes, the watch was subdivided into two quarters, each under the immediate direction of a quartermaster (*kvartermästare*). In naval vessels in the age of sail, the smallest unit was traditionally called a mess, a group of six to eight men who ate and slept together. In ships armed with cannon, this was also the size of a normal gun crew, the men who would serve one gun. There was no formal leader, but one member was usually perceived as the senior person and commanded the respect of the rest.

The sailors were supplemented by specialists with particular tasks. The cook (*kock*) and the steward (*skaffare*) provided meals, while the provost (*profoss*) maintained order and handled minor disciplinary matters. The master gunner (*konstapel*) and his 20 gunners (*bysseskyttar*) directed the soldiers in handling the great guns.

In addition to the men specified for *Vasa*, the ship probably carried her share of the other specialists employed by the navy. A chaplain (*präst*) saw to the spiritual needs of the crew and conducted divine service. In a strongly Lutheran Swedish military, religious services were frequent and attendance enforced. Prayers were held every morning and evening, with sermons on Sundays and holy days. For those sick or wounded in body rather than spirit, rough and ready medical care was provided by a barber-surgeon (*barberare*), who could amputate limbs or cut hair as the occasion demanded. Damage to the ship was repaired by carpenters (*timmerman*) supplied and paid by the shipyard entrepreneur. A ship the size of *Vasa* would typically carry four carpenters, who were technically civilians. The navy supplied musicians, usually a drummer or trumpeter, for transmitting orders in larger ships, and each ship was assigned a few ship's boys for waiting on the officers and running messages. Because *Vasa* was the flagship for the home squadron, the vice admiral in command, Erik Jönsson, was aboard, probably with any staff he had. All told, the population on board if the ship had entered service could have been up to 450 souls, sharing the crowded space of the gundecks and cabins.

The sailing crew took turns on duty, standing watch for four hours at a time. This meant that the sailors could not sleep through the night or for more than four hours at a time. The interrupted sleep schedule was one of the many inconveniences of shipboard life, and modern research shows that it is unhealthy, leading to long-term fatigue, accidents and reduced productivity. One watch was enough men for normal sailing, but not enough for getting under way or battle. In these situations, the watch schedule was suspended and all hands were called. Some of the specialists, such as the carpenters and the cook, did not stand watches but normally worked during the day and slept through the night.

NO ELBOW ROOM

A ship is a community, which must be self-sufficient and self-governing. With such a large number of people crammed into a small area with no chance of escaping, it requires active measures to maintain harmony and a bearable working and living environment. It is not a land-based community in miniature, as some have suggested, since there are rarely women and few children or elderly, but it reflects some aspects of the society from which the crew come.

Time was kept for the crew with sandglasses which lasted 30 minutes, but this high-quality gilt brass table clock, made in Germany, was found in the great cabin. It is one of two mechanical clocks on board, signs of wealth and status but not accurate enough for navigation.

There were multiple tiers of formal authority, enforced by systems of reward and punishment, which were quite harsh compared to modern practice. Many offences were punishable by death, and even seemingly minor infractions resulted in beatings or other physical abuse. Because a ship is its own world, in which emergencies must be handled with the people and resources onboard, discipline and order are necessary for safety, and a breakdown in order threatens everyone. A thief or a shirker was thus not just an annoyance, but a real danger. In the tight confines of the gundecks, small conflicts could quickly escalate to violence. Those in authority kept a careful watch on the mood of the crew, and had to be alert to subtle tensions among the men. In this environment, it was felt that only harsh discipline would work on a crew made up of unwilling conscripts, many of whom may have been undesirables in their home parishes.

One significant factor in the social organization on board was the internal layout of the ship. The division of the interior, the balance between public and private space, and the control of access to particular compartments all affected the way the crew functioned as a community. *Vasa* looks enormous as she stands in the museum today, but the interior was small for all of the functions it had to contain, from storage to sleeping accommodation to battle.

Vasa is essentially a seven-story building, with five levels that run the length of the ship and two smaller "stories" at the stern. The hold, the lowest level,

A FLOATING COMMUNITY   105

A FLOATING COMMUNITY

More than 400 men had to live and work in the confined space of the ship, and the interior was organized to encourage a smoothly functioning shipboard society.

The king's representative lived in the great cabin, or the king himself if he was on board.

Almost half of the crew lived on the upper gundeck, among the 24 cannon mounted here.

Vasa's flags and streamers were probably stored in the tiny coach.

The upper cabin was divided into three rooms and could sleep at least four, in fold-out double beds.

The whipstaff was tended in the steerage, where there were also two berths for pilots.

These sails were not yet bent onto the yards when the ship sailed, but were stored in this carefully built compartment on the orlop.

The orlop, with its low headroom, was primarily a storage space.

Food was prepared in the brick-lined galley.

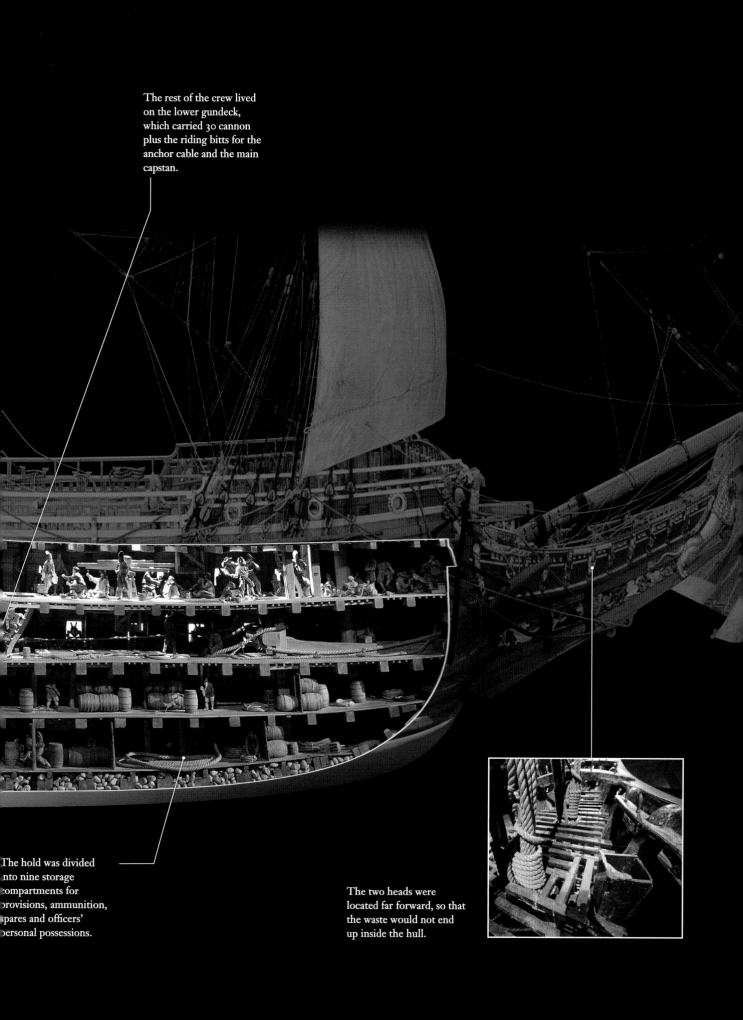

The rest of the crew lived on the lower gundeck, which carried 30 cannon plus the riding bitts for the anchor cable and the main capstan.

The hold was divided into nine storage compartments for provisions, ammunition, spares and officers' personal possessions.

The two heads were located far forward, so that the waste would not end up inside the hull.

Ladders either side of the pump, leading up from the lower gundeck. The pump is a solid alder log over 8 metres long, bored out down the centre.

was primarily a storage area for provisions, ammunition and other heavy items. It was divided into nine compartments by bulkheads with sliding doors. The middle compartment, just before the mainmast, contained the galley, where meals were prepared. This was a large brick structure built on a floor of clay and sand, suspended above the bottom of the hold. This compartment also had about half of the ship's round shot for the 24-pounder cannon. Provisions were kept near the galley. Just forward was a small compartment completely full of barrels of meat, while water and beer were probably kept in the compartment just aft.

Farther forward was a large compartment mostly devoted to storing the anchor cables and mooring lines. This also contained the rest of the 24-pounder round shot and a large quantity of spare parts for the rigging. At the stern, there were compartments for storing the personal possessions of the officers. Most of the hold was empty or nearly empty when the ship sailed, since the soldiers had not yet come on board with their equipment and the ship was not provisioned for a long cruise. One of the compartments in the hold should have been the powder magazine, but none of the compartments has the special construction normally used for storing gunpowder, and it is not yet clear where the gunpowder recorded as taken on board in July, 1628 was stored.

The orlop, immediately above the hold, was also used for storage. This space was very low, with barely a metre of headroom under the beams, and must have been very difficult to move around in. It was mostly one large compartment, with smaller rooms at the ends of the ship. The large compartment contained everything imaginable, from anchor cable, raw lumber and lanterns all the way up to a spare tiller over 9 metres long, which is still stored on board.

At the bow was a small compartment with some of the gunners' stores and other shipboard equipment. Three compartments are arranged at the stern. The largest of these was constructed in the same way as the gundecks, as two old-style 24-pounders were to be mounted here, firing through the transom. The carriages were in place, but not the guns, when the ship sank. This compartment contained the specialized anti-rigging and anti-personnel shot for the 24-pounders and all of the 3-pounder ammunition, as well as spare trucks (wheels) for the gun carriages. Just forward, on the starboard side, was the carpenters' store, with a chest full of tools and some personal possessions. To port was a very carefully constructed compartment with a hinged door and latch. The six sails not yet mounted on the

yards when the ship sailed were stored here, together with spare parts for the pumps and some other rigging equipment.

The gundecks were designed as large open spaces with minimal obstructions, so that the guns could be moved and aimed as easily as possible. All of the things that might impede movement, such as the masts, capstans and ladders, are concentrated along the centreline. On the lower gundeck, this includes hatches for access to the orlop and hold, while the upper gundeck has a nearly continuous series of gratings to provide light and air to the lower gundeck. The after end of the lower gundeck was closed off by a bulkhead with a sliding door to create the gunroom, a compartment for the storage of gunners' equipment and where the tiller swept back and forth. Four guns would have been mounted in this room, one on each side and two in the transom, but the transom guns had not yet been delivered when the ship sailed. All told, thirty guns could be mounted on the lower gundeck.

The upper gundeck had only 24 gun positions, since the after third of the deck was taken up by the steerage and the great cabin. The gundecks were, in

The lower gundeck, looking forward towards the massive riding bitts for the anchor cables. It is a large, open space now, but was originally subdivided into smaller spaces by the cannon at the gunports. The sailors and soldiers slept here, on the deck between the guns.

The helmsman's position, as seen from the doorway to the upper deck. The space is very high to accommodate the long whipstaff, and the bench on which the helmsman stands was added as an afterthought, probably once it was discovered that the deck was too low to allow a view onto the upper deck.

practice, divided up into smaller spaces by the cannon at the gunports. These defined "rooms" about two metres long, each of which was shared by a mess. The men slept directly on the deck or on simple pads, similar to coarse shag rugs which could be rolled up and stowed easily.

The steerage was a tall, narrow space, to make room for the long whipstaff, with doorways into the gundeck and ladders up to the upper deck. A sleeping berth with a sliding door was built into each side, probably for the pilots. The great cabin, reached by a door from the steerage, was the seat of power on board the ship. A carved lintel with a crown supported by tritons and sculpted herms to either side of the door made the message clear: those who live here wield the king's authority. In normal use, it was occupied by the admiral and others of high rank. It was richly fitted out, with fine panel work, a false floor of wide, pine boards and painted sculptures, as if it were part of the royal palace. A bench ran along all four sides of the room, concealing four fold-down double beds. Even the king shared accommodation in the cramped confines of a warship, if he was on board. The great cabin had the only movable furniture found in the ship, a table and chairs as well as free-standing cupboards. A bulkhead with a glazed window separated the cabin from a small, enclosed stern gallery with tall glass windows. The stern gallery provided access to the quarter galleries on either side and, via a staircase rather than a ladder, the upper cabin above. There were four small round ports and mounting points for swivel guns in the stern gallery, and each quarter gallery had a small round gunport at its forward end, suitable for a musket, but there was no provision for any heavy armament in this part of the ship.

The upper cabin was a lower and plainer space, divided into three rooms. In addition to the staircase from below, it could be reached by two doors and short ladders down from the small deck atop the steerage. The forward room, through which the mizzenmast passed, was very narrow and looks to be the unplanned result of moving the mizzenmast forward during the construction of the hull. There were two fixed benches with fold-down double beds in the central room, and the after room gave access to the tiny upper quarter galleries. Four small windows in the transom and one each side in the central cabin, all closed with wooden shutters, provided light and air. Although not nearly as opulent as the great cabin, it was still a comfortable space with adequate room for four people.

The captain probably lived here, as there was easy access to the quarterdeck, and he may have shared the accommodation with the lieutenants or the master.

The communal living space for officers was different from what was found on French or English ships, with their multiple small cabins. This is partly the result of the less harshly stratified social order in Sweden, which lacked the carefully ranked tiers of landed aristocracy of the feudal monarchies to the south. There were relatively few nobles, and the fiscal-military state valued ability over birth. Even though the king had promised, in the Form of Government he agreed with the nobles on his accession in 1611, that all higher offices would be reserved for nobles, all of *Vasa*'s officers, even Vice Admiral Erik Jönsson, were commoners. The arrangement was probably also a reflection of the style of leadership encouraged by the king. Gustav Adolf was a warrior king who enjoyed the soldier's life. In the field, he made a point of sharing some of the hardships of his men and encouraged his generals to do the same, earning the loyalty of their men. He led his army from the front, a rare quality by the 17th century, and was one of the few European kings after 1600 to die in battle, leading a cavalry charge at the Battle of Lützen in 1632.

This carving of a crown held by tritons, mounted over the door into the great cabin, informed the visitor that the king's authority resided in this room.

The upper and lower quarter galleries were a distinctive feature of this period, but seem to be mostly ornamental. There is little space inside, and the upper galleries were so cramped that it was not possible to stand upright. In later ships, the officers' toilets were located in the galleries, but there is no sign of them on *Vasa*.

The coach, the uppermost cabin under the poop deck, was low and narrow, with no windows and a single door opening directly onto the quarter deck. The room's function is not clear, although later ships often stored flags and streamers in this location.

The weather decks, those open to the sky, were working spaces and had to be kept clear when the ship was sailing. The upper deck had gratings down the centre to provide light and air to the gundecks. The deck was narrow, as the sides of the ship leaned inward above the waterline, and the 3-pounders and assault guns carried on this deck impeded movement. The 3-pounders were so long that they could not be withdrawn all the way from the gunports in most locations, so men would have had to stand on the chainwales, exposed to enemy fire on the outside of the ship, to load them if *Vasa* had ever seen action.

A FLOATING COMMUNITY   111

The raised decks in the stern provided the officers in command of the ship with a good view of the deck, the rig and the surrounding water. Although there was little space, many of the sails of the mainmast and mizzenmast were trimmed from the sterncastle. These decks were also useful in battle, as they allowed men with muskets to fire down into lower ships from behind the protection of the bulwarks. Men on deck were further obscured from the enemy by arming cloths, painted canvas hung from the uppermost railing to close the open space beneath.

The final working space in the hull was the beakhead, the long projecting structure at the bow. The rigging of the sails set on the bowsprit, the spritsail and spritsail topsail, was handled in the narrow, triangular space defined by the head railings either side. It was floored with a grating instead of a conventional deck, which allowed water to drain away quickly. The ship's heads (toilets) were also located in the beakhead, a pair of wooden chutes with classic outhouse-style seats. Two heads do not seem like enough for 450 men, but it was also possible to stand or squat on the chainwales on the sides of the hull. The officers living in the cabins at the stern had pewter chamberpots for their sanitary needs.

BAD BACKS AND BROKEN BONES

Vasa's crew have left only faint traces in the historical record, but their possessions remained aboard when the ship sank, and an unfortunate few went to the bottom with the ship. We do not know exactly how many people were on board when *Vasa* sailed in 1628, just that the soldiers had not yet been embarked and the mariners were allowed guests for the first leg of the passage out to the Archipelago (they were to go ashore at the first night's anchorage, the fortress at Vaxholm). Contemporary accounts of the maiden voyage note that women and children were on board. It was not unknown for families to accompany crew when ships were in home waters, and the maiden voyage of a new ship was a festive occasion in which guests took part. All told, there were probably over 150 people on board, and there may have been as many as 200.

Fifteen skeletons and a number of other bones were found when the ship was raised, together with thousands of personal objects. These remains and the chests and barrels of clothing, shoes, games, and money tell us much about the people of *Vasa*, even if we cannot know their names. The skeletons are of thirteen men and two women, ranging in age from late teens to perhaps over 60. The bones preserve much of the medical history of the crew, in addition to basic size and age information. The average height of the men is 1.67 metres, considerably shorter than the current Swedish average of 1.80 metres. The shortest man in the group was only 1.60 metres tall, while the tallest was about 1.80 metres. Body type or build covered a similarly wide range, from very thin to muscular.

Age is harder to determine for adults than height, but it is possible to say that the *Vasa* sailors spanned the full range of working adulthood. Statistical data for life expectancy only go back to the 1730s in Sweden, but it is thought that the data for the 17th century are similar to those a century later. Life expectancy at birth was only about 33 years, because most people died before their tenth birthdays. A man who made it to 20 years old had to have an excellent immune system and so a reasonable chance of making 60. Thus it is not surprising to find crewmen who are certainly in their 40s and one or two in their 50s, although most are probably under 40. The men show generally good health and nutrition, although a higher incidence of healed broken bones than the modern Swedish population in general.

Contemporary illustration of a peasant from Eastern Bothnia (in modern Finland), wearing the same clothes as found among *Vasa*'s sailors: doublet, loose breeches, stockings and a woollen cap.

The two women are a very different story. One is in her late teens, and suffers from a degenerative disease of the spine, to the point that she was a hunchback. The other, in her 20s, suffered from malnutrition or severe disease as a child. Chemical analysis of the bones shows that both women were suffering from anaemia or chronic diarrhoea when they died, either of which could be related to poor diet. They share some similar rare, hereditary skull features, so it was once thought that they might be sisters, but mitochondrial DNA analysis has proven that they had different mothers. DNA has been successfully extracted from ten of the skeletons, and none of them appear to be related maternally. Paternity cannot be determined from mitochondrial DNA, which is what survives in archaeological skeletons.

The skeletons were given letter designations when they were found, later expanded into names to make them more human. Skeleton A became "Adam," skeleton B became "Beata," C became "Cesar," etc. It allows us to see them as people rather than artefacts. Modern forensic sculpting techniques also make it possible to give them faces, based on their preserved skulls, average soft tissue thicknesses and knowledge of their age and health. These reconstructions are not fantasies, and their use in modern murder cases shows that a person who knew one of *Vasa*'s crewmen in life would probably recognize him from the reconstruction.

"Adam" was the first skeleton found and is a good example of a typical sailor. He was in his 30s or early 40s when he died, about 1.72 metres tall, of average build and in generally good health. He had a slight stoop and a large number of healed broken bones. In addition to broken ribs, at some point in his life he had been struck in the face by a heavy object, which broke his nose, cheek and the ridge above his right eye. None of these was properly set, so his face is asymmetrical, with a visible dent above his eye and a nose which points off to the left. His clothing included the baggy, knee-length trousers and short-skirted jacket typical of farmers, fishermen and manual labourers in the later sixteenth and early 17th century all over northern Europe. His were made of undyed, coarse wool, and the jacket fastened with hooks and eyes of brass wire. His shoes were of stout leather, without heels. Heeled shoes were a fashion that came into Sweden in the 1610s, but were more expensive than the traditional flat-soled shoe worn by most people. His knife sheath was of wood, with several sets of initials and possibly a date carved in it, suggesting that he was literate. Fragments of a small book found near him may also be his, although there is no information that identifies what the contents were.

"Filip" may have been steering the ship when it sank. He was a tiny man, only 1.63 metres tall and very thin – he probably weighed less than 45 kg. Isotope analysis of his bones shows that he ate almost no meat or fish and got his protein from plants, which would explain his thin bones and light musculature. He was probably not a vegetarian in the modern sense, but was reduced to a diet of porridge by his teeth. He had supernumerary teeth in his lower jaw, giving him a double row of incisors, which ruined his bite. When he closed his mouth, his molars did not meet, so it was probably difficult to chew solid food effectively. He was dressed in much the same way as Adam, although he had spent more money on his jacket, which was closed with a dozen buttons made of black glass.

"Ivar" was a mature man, possibly in his 50s, over 1.70 metres tall and powerfully built with good general health, although he had his share of healed injuries. The most dramatic of these was a depressed skull fracture, on the left side of the back of his head. It now appears as two small round depressions either side of a mass of scarred bone, and must have been a frightful injury. His clothes

The shoulder wings are a simplified version of a decorative detail seen on more expensive clothes.

The front could be closed by buttons, but many jackets (including this one) had only wire hooks and eyes.

The short skirts were common on the jackets of peasants. Higher status jackets had long skirts, reaching to mid-thigh.

A jacket of dark blue wool, found in a sailor's chest on the upper gundeck. It is a typical working man's doublet, common all over northern Europe in the 16th and 17th centuries.

did not survive in identifiable form, but his jacket had a silver button on it, and he was wearing gloves. A military issue musket and a wooden lantern were found near him, and he may have been carrying them shortly before he died.

The skeleton designated "Johan" may be the one person who can be identified by name. He is an older man with heavily worn teeth. He had broken his left shinbone at some point, but it had not healed correctly, either because it was not set properly or he started walking on it too soon. He had also lost his right big toe in some sort of crushing injury, which would have given him a pronounced limp. He was wearing a long-skirted jacket of fine, black-dyed wool twill, very stylish for 1628.

Who might an older, lame man in a fine jacket be? Certainly no ordinary seaman. Of those who died in the sinking, about half are represented by the skeletons. The only one of the dead mentioned by name in the sources is Old Captain Hans Jonsson, who was trapped below decks. In theory, his skeleton should be one of those found in the ship. Johan and Ivar are the only two who could be considered "old," and Johan is clearly not one of the crew. Johan is dressed like an officer, in finer clothing. Might he be Hans Jonsson?

Sailors brought their possessions aboard with them in chests, barrels or sacks. Some of the chests are finely made pieces of furniture, with dovetailed corners

The pewter plate, the possession of one of the officers, is inscribed with a complex monogram, while the wooden bowl has a geometric figure.

One officer had a silver spoon, while the ordinary crew had wood. 83 spoons were found in the ship, indicating the minimum number of crew onboard when the ship sailed.

Many personal possessions were marked, either with initials or geometric figures carved into them. Initials suggest that the owner could read and write, while geometric figures were traditional symbols handed down through many generations of the same family.

and interior dividers, while others are rough boxes nailed together from plain boards. The contents are broadly similar, usually consisting of spare clothes and shoes, sewing and shoe repair kits, and small containers of personal foods or medicines, but each is unique. A handful had clay tobacco pipes, but smoking had not yet caught on in Sweden to the degree it would in the next generation. A fair number had shoe lasts with them, either for repairing their shoes or to give to a cordwainer if they needed new shoes. One sailor had two pairs of shoes and a very fine broad-brimmed hat in his chest, found just as he had packed them in 1628.

Many brought their money on board with them. The copper coinage introduced in 1624 was of little value, so a small amount of money was still a large number of coins. Most of the bodies have coins associated with them, and two had over thirty coins each, more than 700 grams of copper. Some of the chests have veritable hoards of copper, with the occasional silver coin. The largest single find was over 900 copper coins, and all told, more than 4,000 copper coins were found in the ship. These have a total value of less than 500 dalers, about the equivalent of the annual pay of ten sailors. Despite the wild speculation that accompanies the discovery of any shipwreck, there was no pay chest or treasure found on board, just the pitiful savings of the men conscripted into the crew.

DULL FOOD AND PRECARIOUS HEALTH

The skeletons show that while the men in the crew led relatively rough, physical lives, they were in generally good health and reasonably well fed. None shows evidence of current dietary deficiency or the scarring that some types of vitamin deficiencies, such as scurvy and rickets, can leave on bone if severe enough. Provisioning records indicate that the navy fed its men a solid if dull diet with adequate protein and carbohydrates, and a decent supply of vitamins as long as fresh food was available. However, the Crown only supplied provisions when a ship was outside of home waters. Within Sweden, the men had to feed themselves. Correspondence between the vice admiral in charge of the Gdańsk blockade in 1628, Henrik Fleming, and the admiralty shows that the navy was aware of the dangers of scurvy (caused by a lack of vitamin C) and how to prevent it, even if the actual cause was not known until the early 20th century. Scurvy was not normally a problem in the fleet when in home waters, since fresh food was readily available, but it did afflict crews on long-term blockade duty on the Polish coast, as did other nutritional deficiencies.

The main sources of carbohydrates were bread and peas, the latter shipped dry in casks and made into thick pea soup and stews. Together these provided plenty of fuel for men engaged in periodically heavy work, although ships often ran out of bread. Protein was also more than adequate, with a regular ration of meat and fish. Analysis of the bones found in the casks in the meat storage compartment shows a surprising variety. Beef is the most common, followed by pork and some mutton, but there is also venison, reindeer, and even a moose. Fowl are represented by chicken, duck, goose and capercaillie (a type of grouse), and there are six species of fish. These include not only herring, the stereotypical staple of Scandinavian cuisine, but also pike, cod, and ling. Some of the more uncommon species may be provisions purchased by individuals for their own use, or animals taken by those who had hunting weapons. The only live animals were two cats, one of them so large its bones were initially mistaken for those of a dog. They may explain the lack of rat bones.

Officers drank from pewter tankards or ceramic jugs such as this Rhenish stoneware example with a pewter lid, while the crew shared wooden tankards.

All of the meat had been cut into pieces that would fit into casks. The casks used on board were in standard sizes. The most common was made of oak and held 125–130 litres, which was one *tunna* (barrel), the standard unit of volume measure for fluids in Sweden in the 17th century. These casks were used for

A FLOATING COMMUNITY 117

In addition to a large cauldron found in the galley, there were at least ten small clay pipkins on board. These could be placed directly in the galley fire to cook a smaller portion of soup or stew.

The galley, seen here from above during the excavation, had a brick floor over a layer of clay and sand, and brick walls on two sides. Several fires could be lit on it at once, although there is little evidence that it was used before the ship sank.

everything that could be put into them, both wet and dry, and some were made with removal lids to carry personal possessions. Some of the compartments in the hold, including the large one behind the mainmast which probably held water and beer, are not accessible from above by hatches, so provisions stored there had to be lowered down into an adjacent compartment and rolled or lifted through the doors to be stowed. For casks weighing 125 kg, this was no small task.

Food was prepared in the galley, on open fires lit on the floor of the brick hearth. Small pots were used in some cases, large cauldrons in others. The largest cauldron found in the ship was in cast iron, with a capacity of about 140 litres, which could feed the mariners in one sitting but not the soldiers as well. Most of the food prepared was in the form of stews, which was dished into covered wooden buckets, called mess kids, for transfer to the gundecks. There, it was ladled or poured into common wooden bowls, one or two for each mess. Wooden plates might also be used, but these are small and probably belonged to individuals. Each man had his own spoon, and the surviving spoons show a great deal of individuality, with painted and carved decoration. Beer or water was served in a wooden tankard, one for each mess, passed around from man to man. Some men carried their spoons with them, but one mess on board *Vasa* had stored all of its eating utensils together, in a barrel on the orlop. This contained two wooden bowls and a tankard, with seven wooden spoons in the tankard.

Although scurvy and dietary deficiencies were not normally a problem in the fleet, disease was an ever-present threat in the close confines of a ship. It was virtually impossible to stop the spread of a communicable disease, even if the knowledge of germs and infection had been available, and disease could quickly render a ship defenceless. Typhoid and other infectious diseases could be deadly, and potentially fatal pathogens such as cholera or dysentery, carried in spoiled food or water, would strike many at once.

Medical knowledge of how these diseases worked was limited, and there were in practice no cures. The symptoms might be treated, and a wide range of pharmaceutical compounds were in use in the 17th century for treating minor and major ailments, but few if any ships in the Swedish navy carried physicians with academic knowledge. It was only in the 1620s that the navy began to require ships to carry a barber-surgeon. He had practical experience of treating the common complaints afflicting people of the day, from colds to diarrhea, as well as serious diseases, and this probably made up the bulk of his work. He was also expected to treat the kinds of injuries common on warships, such as broken bones, open wounds and gunshot. He could set broken bones or amputate limbs and sew up splinter wounds, but the results were not likely to be pretty and infection was a common and potentially fatal development. Analysis of some of the known remedies from the period shows that many included ingredients that had some effect, although there were many that were little more than placebos and some that were actually harmful. There was some understanding of the importance of hygiene, and keeping the ship clean was a normal part of shipboard routine. Personal hygiene was also a part of Swedish and Finnish culture, with a tradition of regular bathing in purpose-built bath houses, a tradition that survives in the Finnish *sauna* and Swedish *bastu*. Even the navy yard had a bath house, with two men paid to look after it.

A matched pair of pewter flasks were found in a fitted wooden case. They may have held wine or spirits. Another pewter canister contained rum, which was still 33 per cent alcohol when it was found in 1961.

These men, their possessions and provisions, came aboard *Vasa* starting in the spring of 1628. Some helped rig the ship, others began to learn the ropes, finding their way among the forest of rigging. Others helped to stow the provisions and equipment coming on board for the maiden voyage. Tension probably increased in that summer, as the ship's departure continued to be delayed. Desertion may have been a problem, with the ship stuck in the town and no action imminent. Fear may have begun to establish a foothold, since many of the men must have been present for the demonstration of the ship's lack of stability in July, arranged by Captain Söfring Hansson for the benefit of Vice Admiral Fleming. Eventually, the king's insistence that the ship get to sea could not be ignored, even if the foundry was still not finished with the armament. In early August, the last preparations were made, and *Vasa* was ready to sail.

Sinking

1628

Captain Söfring Hansson looked over the side to make sure that all of the mooring lines were clear, and was pleased to see the stern swinging out into the harbour. As he turned back to the quarterdeck, his attention was briefly caught by the crowds lining the quay in front of the palace. It seemed like all of Stockholm had turned out to see the ship off. It was warm and almost still on this 10th of August, but that was not the only reason he was sweating. He called down to the main deck, "Petter, haul us ahead!"

The lieutenant acknowledged the order and called down through the hatches to the men manning the main capstan on the lower gundeck, "Heave around!" The men stamped their feet and began to push against the capstan bars, drawing in the mooring cable running out through the hawsehole forward. The end of the cable was looped over one of the wooden pilings farther south, along the waterfront. As the men walked round the capstan, 1200 tons of hull, rigging, guns, men and provisions began to creep forward, towards the far end of the city. There was barely any wind to oppose them, but it was out of the south-southwest, so they would have to warp the ship to the open water at the train oil huts before they could set any sails. This was slow, hard work, taking a line to each piling in turn and hauling the ship up to it with the capstan.

As the ship crept along, people on the waterfront followed, easily keeping pace. They waved and called to their friends on board, teasing them over how long it had taken to get the ship ready and wishing them well. Captain Söfring kept his eyes on the sky to the south, looking for changes in the clouds or evidence of gusts and willing the wind to stay low. As they neared the end of the town island, where one of the outflows from Lake Mälaren entered the harbour, Hansson ordered men aloft to loosen the sails. There were only four sails bent onto the yards, the fore course and topsail, the main topsail and the mizzen, which were all he would need on the tricky passage through the archipelago to the summer fleet base on Älvsnabben to the southeast. With this light breeze on

The anchor was raised with the capstan on the lower gundeck.

The officer of the deck would have stood here, where he could see most of the upper deck and the sails, and where he could give orders to the helmsman. If he needed a better view, he would climb up to the quarter deck.

121

the beam, he would need all four: the course and topsails for power, the mizzen to balance the helm. In the narrow confines of the harbour, with a lee shore only a few hundred metres to the north, he needed all the steering control he could get. He knew that he could ride the current from Mälaren through the narrowest part of the upper harbour, where he was in the lee of the bluffs of Södermalm, but he would not be able to steer properly until he had some wind in the sails.

At the last piling, the men looped a second line over it from one of the stern hawseholes and took in the warping line. The current carried the bow east while the remaining line held the stern up to the west, turning the ship to bring the wind on the beam. Once the ship was pointed in the right direction, the stern line was cast off and the ship began to move eastward. "Sheet home topsails and set the main topsail!" called the captain. The master, Jöran Matsson, and his mate already had the men in place on deck and aloft and started issuing the detailed commands that freed the topsails from the gaskets and sheeted home the clews, the lower corners of the sails. With these belayed, a large gang began hauling the halliard to raise the yard and stretch the topsail. It did not fill immediately, and the ship drifted down the harbour with the current, yawing slightly from side to side in the eddies.

Men sent aloft to loose the sails climbed the ratlines, ropes tied across the shrouds like the rungs of a ladder.

"Master gunner, the salute, if you please." The master gunner nodded, and men standing by guns loaded with powder and wadding but no ball touched their glowing matches to the priming pans. The pans hissed with fire and smoke before a great, thunderous roar belched from the muzzles of the bronze cannon, ringing and echoing from the rocks and buildings. The ship disappeared in a cloud of sulphurous smoke, hanging in the nearly still air, to reappear as the current carried her clear.

"Set course and mizzen!" Men scurried to new positions to follow the captain's order, encouraged and guided by the lieutenants and the master. The fore course, a mass of hempen canvas, fell from the yard, but there was not enough wind to pull the sheets through their blocks, even though they were well greased. Master Jöran sent men to pull the sheets through by hand, so the sail would hang properly.

As they were setting sail, a puff of wind came down off the cliffs and the topsail briefly filled. The ship heeled, but slowly righted herself. Söfring let out the

breath he had not realized he was holding. He checked the men and passengers, wives and children along for a nice afternoon sail down to the fortress at Vaxholm, where they would disembark once the ship anchored for the night. The new men had little experience of the sea and every movement of the ship startled them, but he could see the older men, with hands and souls calloused from years of service, exchanging dark glances and looking anxiously to the south.

    The captain watched the water surface, looking for ripples, signs of a gust or freshening breeze. Like every sailor who had taken a ship out of this harbour in a southerly wind, he knew there would be a sudden increase in wind force as they sailed past the gap in the cliffs at Tegelviken, where the wind from the high ground was funnelled down to the water. That spot would be the first test. If he could not make it past there, he had no chance of getting the ship all the way to Älvsnabben in the southern archipelago. If he could reach Älvsnabben, he might have time there to find a remedy for the ship's atrocious lack of stability. He knew the ship was crank, heeling far over in the slightest zephyr, but until he had sailed her some distance he would not know enough about her behaviour to be able to find a working compromise of ballast, stowage and sail area. He was glad

A ship leaving Stockholm, painted by the Dutch artist Bonaventura Petersz in the 1630s. The view is to the north, with Slussen in the foreground and the city centre (Gamla Stan) behind. The ship is using the same four sails *Vasa* had bent to the yards in 1628.

Vasa is towed round to the Tre Kronor palace, which was also the arsenal, to take on her armament, spring 1628.

Vasa is launched at the navy yard, spring 1627.

Skeppsholmen (Blasieholmen)

Lake Mälaren

Stadsholmen (Gamla stan)

The train oil huts (where seal oil was stored), now Slussen, with a permanent current from Lake Mälaren.

Vasa casts off the warp, and begins to set sail, main topsail first, then fore course and mizzen. She heels for the first time.

10 AUGUST

A PUBLIC CATASTROPHE

Vasa's entire career covered little more than a nautical mile, from launching in 1627 to sinking a year later. The ship managed to set three sails and was just managing steerage way on from the southwesterly wind, when she heeled over until the gunports were under the water.

Vasa casts off from the palace and is towed along the waterfront of Stadsholmen, between 4 and 5 o'clock in the afternoon of 10 August, 1628.

Today, the heights of Södermalm form a vertical rampart on the south side of the Stockholm harbour. In the 17th century, they were still rounded from glacial scouring. At Tegelviken, a pass channels the wind down to the water.

(Skeppsholmen)

(Södra Djurgården)

Vasa uses a stern line to turn to the east.

(Kastellholmen)

Strömmen
(Baltic Sea)

Biskopsholmen
(Beckholmen)

As *Vasa* passes the end of the cliffs at Tegelviken, a gust heels her so far over that she begins to fill with water, and sinks.

The ship drifts with the current until enough wind fills the sails and she begins to answer the helm.

Tegelviken

People abandon ship as *Vasa* heels over and begins to sink. Small craft following the ship rush in to rescue the victims.

that they were still eight guns short of the ship's full armament. It was another 75 *skeppund* of weight high up in the ship he could do without.

As they passed the end of the cliffs, a second, stronger gust filled the sails. He knew as soon as the ship started to lean that she was not going to stop. The wind would press her over and down until the open gunports were in the water. "Cast off the sheets! Helm hard to leeward! Haul in the guns and close the ports!" If they could loosen the sails and turn up into the wind, dumping the gust out of the sails, the ship might right herself before she had taken on too much water. It was risky, since the ship would pass through a point where the force on the ship was even greater, but the alternative was to turn away from the wind, which would take them straight onto the rocky shore of Beckholmen. If they could get the guns in and close the port lids, that would keep the water out. Söfring looked round for Vice Admiral Erik Jönsson, who had rushed below at the first gust to make sure the guns were secure – if one of them pulled loose and rolled to the low port side, there would be no saving her – but he must still be below, God help him. He could not see old Hans Jonsson either, and with his lame foot he would have a hard time moving around, but at least he knew what to do.

The ship heeled farther, well down on the port side, and he could feel her stagger as the water began to pour in through the gunports. It was all happening slowly, even gracefully, but the captain knew it was too late. He could not save the ship, but he could still save the crew and passengers. He looked to port, where the shore of the island of Beckholmen seemed so close he could almost touch it.

His attention was yanked away by the screams of the passengers. They had been startled by the first gust but laughed it off nervously, embarrassed by their own lack of experience. Now they knew something was wrong and panic had set in, people climbing over each other as they tried to reach the higher starboard side and find something to hold onto. How many were still below decks? It could not be easy to get up the ladders now they were heeled so far over, and he could not see anyone emerging onto the deck. He wondered how many could swim, and looked astern. The ship's longboat was useless, it was being towed alongside and had already been pushed under. The gig towed astern would be dragged down before it could be loosened. A large number of small boats had followed them down the harbour, and he hoped they were still following. They were, good, they would be needed in a few minutes.

The ship was just starting to turn to starboard when it came to a stop and began settling – the water must be over the hatch coamings and pouring down into the hold. As more and more of the gunports sank into the water, the torrent grew, pushing the ship down and down. It was starting to be difficult to keep his feet, and the high sterncastle where he stood leaned far out over the water – it would be a long drop if he jumped or fell. He climbed over the rail and stood on the outside of the hull, hoping he could hold on to the rail until he could simply step off into the water. Söfring yelled down onto the main deck, trying to get people to keep clear of the rigging and wait for the rail to go under. Many were already in the water, some floundering while others swam for the shore.

The main deck was now awash, and air was boiling out of the hatches. It was agonizing watching his ship die, knowing there was nothing he could do to stop it. It was the second ship he had lost, an experience he had hoped never to repeat. Would he still be a captain after this?

At last, only the quarterdeck and poop were still above water. He stayed until the last possible moment, then jumped, trying to keep clear of the mizzen rigging. He left it too late, and was pulled under, into the cold, black water. He struggled for what seemed an age, his heavy woollen clothes dragging him down, but at last made the surface, light-headed and disoriented. The water was swarming with frightened people, but the boats were swooping in, their crews shipping oars and reaching over the sides to grasp hands or jackets and haul in sputtering sailors and guests. Some were clinging to the rigging, and were sometimes dragged under to pop up again, or not. The masts were slowly disappearing into the water, still leaning to port, but then they stopped. The ship must have hit the bottom, 18 fathoms below. At least those who could not swim could hang on until a boat could find them. Others were already huddled, dripping in the late afternoon sunshine, on the shore of Beckholmen. Once he was sure that the survivors had been rescued, he ordered a boat to row him back up to town, to the palace, where he knew he would have some difficult questions to answer.

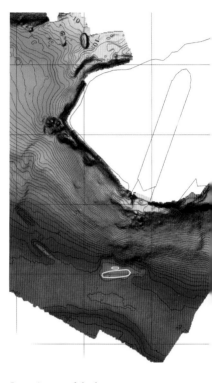

Sonar image of the bottom of Stockholm harbour south of Beckholmen, showing the location where *Vasa* sank.

MISSING, PRESUMED DROWNED

Söfring Hansson was arrested virtually as soon as he reappeared and questioned by the Royal Council the next day. They sent a report to the king in Prussia the following day, informing him of the loss. It took little imagination to see what

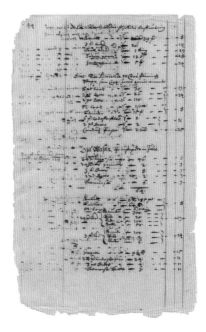

An inventory of the guns and ammunition taken on board *Vasa* in July, 1628.

his response was likely to be, and there was a flurry of activity in Stockholm to try to minimize the damage. The rest of the reserve squadron scheduled to sail to Älvsnabben with *Vasa*, including the ships *Kronan* and *Svärdet*, was held in Stockholm. On the 13th of August, the Council granted a privilege of salvage to an Englishman, Ian Bulmer, who claimed that he could raise the ship.

Bulmer set to work immediately. The ship had come to rest on the bottom, leaning to port 20–25 degrees, with the upper ends of the topmasts still above the water. Bulmer was able to use these to haul the ship upright, probably by attaching tackles to them from ships anchored to the south. He was able to do this very quickly, and there was hope that the ship could be recovered soon after.

Others began to reckon the cost of the loss in both material and human terms. An essential question to be answered was how many had died in the sinking, and who had survived? The first report to the king said simply that the number of casualties was unknown, but it was feared that those who had been below decks, including Captain Hans Jonsson, had been trapped there and perished. Other reports that survive follow a pattern all too familiar from modern disasters. Initially, no one knew anything. As time went on, wild estimates gave way to more careful examination and more accurate numbers. A letter written the week of the sinking by the Danish agent in Stockholm, Krabbe, to King Christian IV in Copenhagen, informed him that "over half a hundred" had died. James Spens, who represented Charles I of England, wrote to London two weeks after the sinking and reported that about 40 people had been lost. At the same time, Chancellor Axel Oxenstierna's brother Gabriel, who was a member of the Royal Council governing Sweden in the king's absence, wrote to Axel with a detailed report of the loss.

Like all of the others, Gabriel Oxenstierna reported both the number of casualties and the number of bronze guns lost. The latter was, strategically, the more important figure and of greater interest to the readers. Gabriel is nearly correct in the number of guns (62; the inventory made before the ship sailed records 64 bronze guns on board), and if the two small 1-pounders are not counted as "real" cannon, then he is exactly correct. He also has the depth of water right (18 fathoms is about 32 metres, which is still the depth of the channel at this spot), and he does not equivocate about the number of casualties. He was in a position to know the results of the muster held after the sinking, so his account is probably the most reliable, at 30 victims, including women and children.

Most of those on board survived, since the ship sank only 120 metres from shore and there were many boats nearby to rescue the victims. Although we do not know what proportion of the population could swim, the fact that so many survived suggests that at least some could swim or tread water. The Council's report, in stating that they thought those trapped in the ship had died, suggests that they expected those not trapped had a good chance of surviving, which implies some sort of swimming ability. In later centuries, it became a tradition that deepwater sailors did not learn to swim, as it was not seen as a useful skill, but for ships sailing near the coast, as ships in the Baltic did, swimming could be useful indeed.

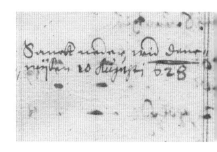

In the margin of the artillery inventory, someone has noted "Sunk at Danneviken 10 August 1628." (Danneviken is near Tegelviken).

The bones found in the excavation of the hull in the 20th century represent at least 17 people (of whom 15 are well-defined skeletons), and thus more than half of the presumed victims. These bones tell us more than just their medical histories, they also provide important clues to what happened on board in those last minutes.

All 15 of the well-defined skeletons were originally very well localized. Three were disturbed during the salvage in the 1950s, but their original positions can be reconstructed with confidence. The lack of disturbance and the presence of clothing and personal effects suggest that they were found near the spots where they died.

Six people were found on the upper gundeck, two on the lower gundeck, one on the orlop and two in the hold. Four more were found outside the ship, in the wreckage of the port quarter galleries at the stern. In addition, a pair of boots with foot bones still in them was found outside the ship at the bow, and a few loose bones that cannot belong to any of the other skeletons were found in the ship. The overall distribution suggests that most people were not very deep in the ship or were making their way upward or outward when the ship sank. Many were found near ladders. One person, "Ludvig," (the names were assigned by the osteologist studying the bones) was halfway through a doorway in the forward bulkhead of the galley compartment in the hold, probably trying to reach the ladder just a couple of metres away that would have led him up to the gundecks. The other person in the hold, "Johan" (the best candidate for Captain Hans Jonsson) may have fallen down the hatches from the lower gundeck, just abaft the riding bitts, and landed on the anchor cable coils. Much of his skeleton was

THE DEAD

Fifteen people, in the form of skeletons, were found in and around the ship during the archaeological excavation, as well other bones that must represent further individuals. Forensic sculptor Oscar Nilsson has been able to reconstruct the faces of several on the basis of their skulls. They were given names by the osteologist who studied them, but all but one are anonymous in reality.

Rudolf, Sigurd, Tore and Ylva: four younger people found in the collapsed wreckage of the port quarter galleries.

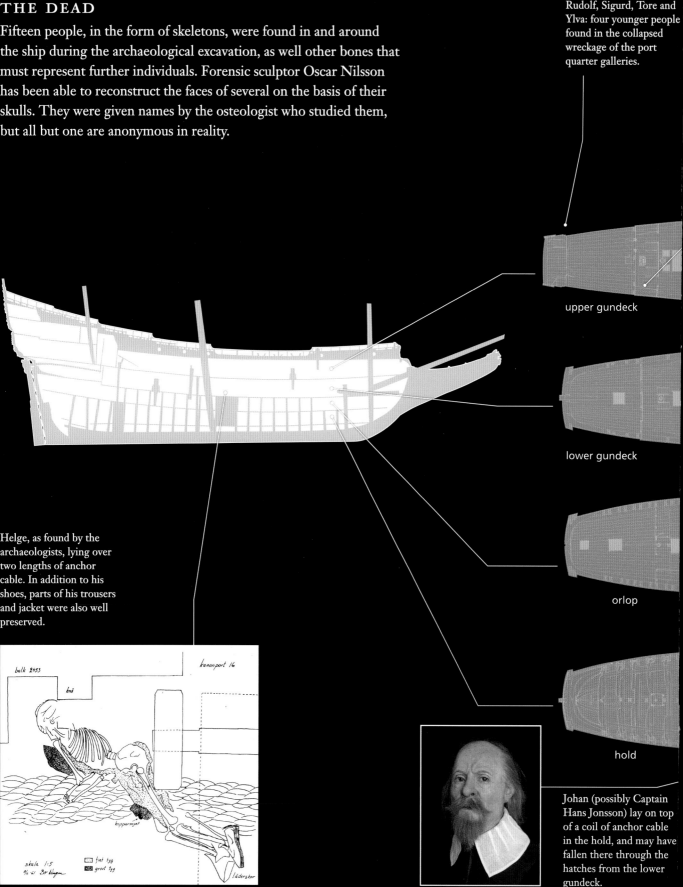

upper gundeck

lower gundeck

orlop

hold

Helge, as found by the archaeologists, lying over two lengths of anchor cable. In addition to his shoes, parts of his trousers and jacket were also well preserved.

Johan (possibly Captain Hans Jonsson) lay on top of a coil of anchor cable in the hold, and may have fallen there through the hatches from the lower gundeck.

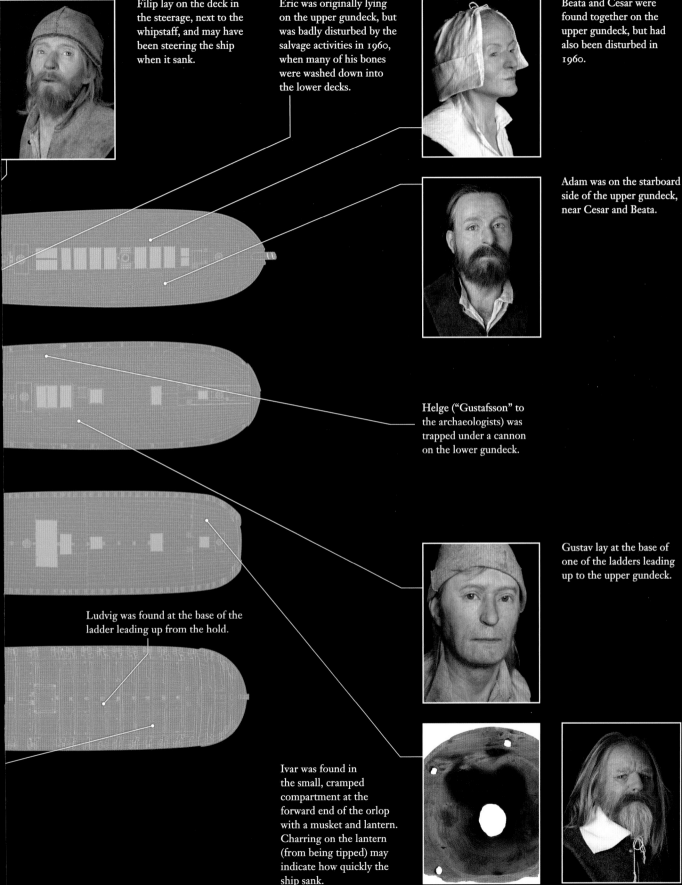

found on top of the starboard coil, where he probably crawled to stay above the rising water.

Two victims suggest action taken at the time of the sinking. "Filip" was found in the steerage, under the end of the whipstaff. He may have been steering the ship when it sank. The rudder was hard over to starboard, turning the ship into the wind, the proper action at the time. He was only a metre from one of the ladders up onto the upper deck and safety. Did he stay at his post, trying to save the ship rather than saving himself? "Helge" (originally called "Gustafsson" by the archaeologists who found him) was trapped under a gun carriage on the lower gundeck, on the port side where the water first rushed into the ship. He was probably one of the first victims. Was he one of those ordered to haul in the guns and close the port lids, or was he already below?

Most of the dead were found alone, but "Beata" and "Cesar" were found together, their bones intermingled on the upper gundeck. Their clothing and possessions indicate similar social status, one of the ordinary seamen and a slightly younger woman. It is tempting to see them as a couple, husband and wife holding onto each other, unable to get up the wildly tilting forward ladder in the rising water.

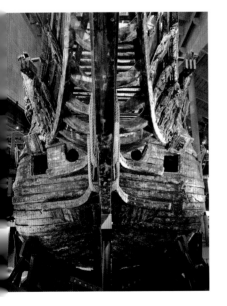

Vasa's bow is of typical Dutch form, flat and rising rather than V-shaped. In the right conditions, she might have been a fast ship for her time.

One lonely man, "Ivar," may hold the best clue to how quickly the ship sank in the lantern he was using when he died, in the small compartment all the way forward on the orlop. The find location suggests that he had set the lantern down on the deck. It has a hole in its lid to allow the heat of the tallow candle inside to escape, but the underside of the lid is charred to one side of the vent, as if the lantern had been tilted to one side about 10–20 degrees, the angle of heel while the ship was filling with water. The lid had just begun to char all the way through to the outer surface when the heat was removed. Could the amount of charring indicate how long the ship heeled before this compartment was flooded, putting out the candle or overturning the lantern? If so, it is a matter of just a few minutes.

WHAT HAPPENED?

The question then and the question now is the same: How did this happen? How had an experienced shipwright and an experienced captain managed to build and sail a ship less than a nautical mile before it sank in a light breeze on a fine summer day? Why did the ship sink?

A well-designed ship should roll back upright when wind or waves push it over. It does this because the forces of weight, pushing down, and buoyancy, pushing up, are balanced when the ship is upright. If the ship heels to one side, buoyancy should increase on the low side, so that buoyancy and weight are offset, causing the ship to roll back up. The farther apart the two forces are, the greater the force resisting the tendency to roll. This force, the righting moment, can be affected in two main ways. The shift of buoyancy to the low side is determined by the hull shape. A hull form that becomes significantly asymmetrical when it heels, creating a large righting moment, is said to have good form stability. The distribution of weight in the ship is the second major factor. The lower the weight is, the greater the offset between weight and buoyancy when the ship heels and the greater the righting moment; the ship is said to have good weight stability. If there is too much weight high up in the ship, not only will the righting moment be smaller, but it can even become negative, causing the ship to roll over, or capsize.

In *Vasa*'s time, few warships had significant form stability. Partly this was the result of the state of naval architecture. The mathematical tools for calculating the stability of a hull form were more than a century in the future, and so there was only a rudimentary understanding of how hull shape affected stability. In this situation, stability was achieved through weight distribution. Shipwrights were careful to build the upper parts of the hull of lighter timbers and the hull was loaded with ballast, dense material such as stone, in the bottom of the hold. The tactical needs of the navy made this a challenge, since effective firepower required many tons of bronze or iron cannon well above the waterline. Multiple gundecks were even more problematical: they required more structure above the waterline, and carried more guns higher up.

Vasa has a fairly typical hull shape for a Dutch warship of the 17th century. It is towards the narrower end of the range, but not unusual. The form stability can be calculated using modern formulae, and these show nothing inherently dangerous about the underwater hull shape. The characteristics are not what we would consider a stable, seaworthy ship today, but very few ships from before 1800 would pass modern stability tests. People of the 17th century accepted a much higher level of risk in their everyday lives than people in the developed world do today. They lived in a world of high infant mortality, lethal diseases,

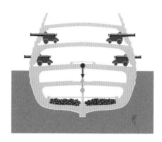
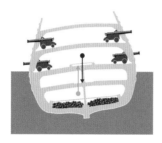

The relationship between buoyancy (blue) and gravity (red) determines a ship's stability. The farther apart the two forces move when the ship heels, the more stable it is. If too much weight is added to one side, such as the water that rushed in through *Vasa*'s open gunports, stability is destroyed.

famine and endemic warfare, an environment modern Europeans would consider frightful, but which they perceived as normal.

Where *Vasa* differed from her contemporaries is in weight distribution. The large number of guns has often been blamed for the ship's instability, but the cannon are not necessarily the problem, at least not directly. The total weight of the guns on board when she sailed was about 62 tons, or about 5.1 per cent of her total displacement (weight of the ship with all guns, equipment and provisions) of just over 1200 tons. This is well within the normal range of wooden sailing ships with two full gundecks, in fact towards the lower end of the range. Even if the eight cannon ordered but not yet delivered by August 1628 had been on board, the total gun weight would still have been in the "safe" range. The problem is in the hull itself. The wooden structure is too heavily built above the waterline and the underwater portion is too small for the amount of hull above the water. The gundecks especially are very heavily built, with beams much larger than needed for the weight of the guns *Vasa* carried and a very complex system of reinforcement. In addition, the headroom in the gundecks is much more than a crew with an average height of 1.67 metres needed. This added extra weight in structure, and placed the weight of the upper gundeck and upper deck, with their guns, farther above the waterline than necessary.

The heavy construction of *Vasa*'s hull helped to preserve the ship, but it also contributed to her doom. The overbuilt decks and their reinforcing timbers, such as these riders in the hold, are a primary factor in the ship's poor distribution of weight.

The problem could not be corrected simply by adding more ballast. This would only have made the ship sit lower in the water, bringing the gunports even closer to the water surface. The lower gunports were not especially low in comparison to later ships with multiple gundecks, but there was not much margin for error. Even seaworthy ships sailed with the lower gunports closed, since normal heeling under sail would bring the ports down to the water. *Vasa* needed to move a significant amount of weight from above the waterline down into the hold in order to be safe to sail. The only effective way to have done this would have been to change the upper gundeck guns for lighter pieces and to place the weight saved in the hold as ballast. Even then, the ship would probably always have been tender and tricky to sail. A more drastic solution would have been to cut the ship down, removing the upper gundeck altogether and rebuilding the upper works. This was not unknown, and with a single gundeck armed with 24-pounders, she would still have been one of the most powerful warships in the Baltic.

FINDING FAULT

It was hard to overstate the scale of the disaster. The newest, largest, most heavily armed ship in the Swedish navy had sunk barely an hour into its maiden voyage, in full view of the population of Stockholm and the agents of the king's allies and enemies. One can only imagine the king's reaction when the Council's letter found him in Poland more than two weeks after the sinking. He wrote back in barely controlled fury, ordering the Council to find out what had happened.

The Council had not waited for a response, but had set in motion an inquiry into the loss. Captain Söfring Hansson was questioned immediately. He swore that the guns were properly secured and the crew were sober, and called upon

At some point during the sinking, the ship heeled far enough to loosen the ballast stones, which cascaded into the port bilge, tearing up the loose flooring laid over them. Once this happened, there was no hope of saving the ship.

Draught marks were cut into the sternpost (seen here) and the stem at half-foot (15cm) intervals, so it is possible to reconstruct how deeply she sat in the water from Petter Gierdsson's testimony at the inquest.

heaven to strike him down if he lied. He provided a brief account of his actions, what sails he had set and what commands he had given, and noted that the ship was too tender. He was released.

The Council called an inquest on September 5th, in the Tre Kronor palace. A tribunal of 17 members, councillors and senior naval officers, sat to hear the testimony of the surviving officers, the shipbuilders, and various experts. The Admiral of the Realm, Karl Karlsson Gyllenhielm, presided and a naval officer acted as interrogator. The transcript survives, although the beginning and end are missing. It begins with Vice Admiral Erik Jönsson, who had nearly been trapped below decks but managed to escape. He was struck by a hatch cover and lay near death for several days after the sinking, but had recovered in time to testify.

Each of the officers was asked if he knew that the ship was unstable before she sailed, and why he had done nothing about it. Much of the questioning focused on ballasting. Erik Jönsson reported that he had gone below to make sure that the guns were secure and gave his opinion about the ship's lack of stability – he believed it would have sunk even if no sails had been set.

Admiral Gyllenhielm interrupted here, and called Erik Jönsson to task. He had been appointed a vice admiral and should have taken care to see that the ship was properly ballasted. Jönsson pleaded the ignorance of the landlubber. "I am neither a captain nor a vice admiral," he said, "but a master of artillery." The tribunal would have to ask the captain or the master about nautical matters. He added that, in any case, more ballast would only have brought the gunports too close to the water. A lack of relevant experience did not, apparently, prevent him from having an opinion (he had, in fact, gone to sea in his youth).

Lieutenant Petter Gierdsson also denied knowing that the ship was crank before it sailed and claimed not to know anything about ballast. When asked if he knew why the ship was so crank, he said he knew nothing of shipbuilding and could not answer. He did notice, before the ship set sail, that people rushing to one side caused the ship to heel. He had been responsible for rigging the ship, and others had taken care of the ballasting. He provided some information about the light winds on the day of the sinking, the sails set and the sailing commands given. He was able to inform the tribunal that the ship drew 14 feet (4.16 m) at the bow and 16 feet (4.75 m) at the stern when she sailed. It is hard to believe that a man entrusted with rigging the ship could not tell that the ship was unstable.

THE INQUEST

A hearing on the loss was held on 5 September, 1628, at which the surviving officers and petty officers, as well as the shipbuilders, were interrogated. Most of the transcript survives and provides a sense of the atmosphere in the hearing room, where a scapegoat was sought and careers were on the line.

The inquest was held at the Tre Kronor palace in Stockholm. This is the outer bailey with the entrance to the council hall.

The next to last page of the transcript summarizes the testimony of various experts.

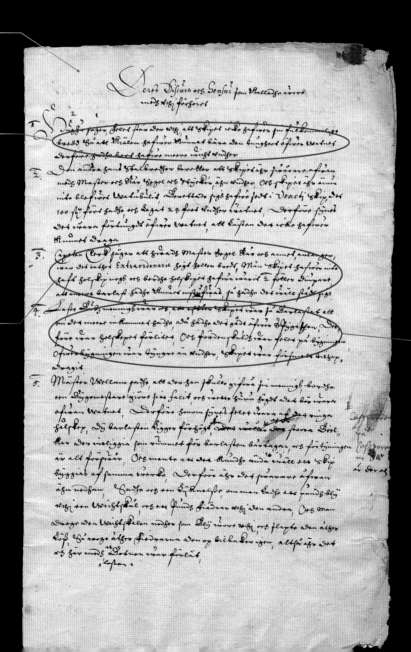

"1. Hughu says that the error is obvious, the ship does not have enough breadth. In order for the keel to bear the weight above the water it should have had more weight beneath."

"3. Captain Clerck says that as far as the masts, sails and yards and other equipment go, it was neither extraordinarily high or wide. But the ship did not have enough hold and should the hold have been 2 feet deeper so that more ballast could have been taken in, it would have stayed upright."

"4. Lasse Bub's opinion was that if the ship had been more deeply ballasted, the water would have come up over the scuppers. There was too little hold at the bow and the ship was improperly built. The upperworks were heavier than under, the ship was too narrow."

A considerable portion of the Swedish state's accounts survive from the 17th century.

The interrogator was not especially hard on the vice admiral or the lieutenant, showing some reluctance to implicate a fellow commissioned officer, but he had his sights firmly fixed on the master, Jöran Matsson. The master was responsible for ballasting, and Jöran willingly admitted this. "I stood in the hold with a lantern, and supervised the placing of every stone. The ballast was properly stowed, and there was not room for one stone more." When asked if he knew that the ship was unstable, he admitted that he did. Aha! "I heard Captain Söfring tell Admiral Fleming that the ship was crank." Jöran Matsson then related the story of the stability demonstration arranged for Admiral Fleming, in which 30 men had run back and forth across the deck to make the ship roll. He told the tribunal that the admiral had ordered the demonstration stopped and said "I wish His Majesty were home." The cat was out of the bag. The admiralty had known in advance that the ship was unsafe.

The master was asked why he thought the ship was so tender, and he replied the she was "too narrow in the bottom" to hold herself upright. He had told Fleming this, but the vice admiral had accused him of putting too much ballast in, so that the gunports were too low to the water. Jöran Matsson answered "God grant that it will still stand on its keel," to which the admiral replied, "The Master Shipwright built the ship well, you should not be worried."

The interrogator accused the boatswain, Per Bertilsson, but his heart was not in it. "You were drunk, like the others!"

"I had taken communion," was the answer. The boatswain agreed with the master, that the ballast was sufficient, and he also claimed not to know that the ship was so dangerous before she sailed. He was dismissed.

At last the shipbuilders came to the stand. After establishing that neither the guns nor the ballast had caused the tragedy, the interrogator took the line that the design or construction of the ship had been flawed from the start. "Why did you build the ship so narrow and badly, that it had no belly and could not stand up, but capsized?" Arendt de Groot and Hein Jakobsson were prepared for this question with very carefully worded answers. They claimed that the ship was as well built as any ever constructed, and many professional shipwrights had said so. Hein Jakobsson said, "I built the ship according to the design given to me by Master Henrik and which the king had approved." Arendt said that he and his brother had shown the king a picture of a similar ship built in Holland, so

that the king knew what sort of ship they would build. Now the interrogator had a serious political problem, how to fix the blame on the builders without implicating the king.

The interrogator tried several different approaches, but was met with the same answer, "We built the ship the king approved."

"Then why did the ship sink?"

"I do not know, only God knows."

The builders knew that to budge from this position was dangerous, but Hein Jakobsson opened the door to a way out just a crack. He said that he had abided by the specification he received, but that he had made the ship 1 foot 5 inches wider; he could do no more due to the advanced stage of construction at the point when he took over responsibility. The implication was clear: Master Henrik may have been wrong, but Hein Jakobsson had done everything he could to fix it. The other master shipwright at the navy yard, Johan Isbrandtsson, was called to the stand, and he confirmed that the ship was very well built. The interrogator asked if the upperworks might have been too heavily timbered in proportion to the lower hull, but Master Johan did not think so. Still, the ship's original design had been implicated. Master Henrik was dead, so could not defend himself and could not be punished – he was the perfect scapegoat.

A nation's collective memory is to a large degree linked to its archives. In the National Archive in Stockholm can be found most of the documents regarding the construction and loss of *Vasa*.

SINKING 139

The transcript ends with a summary of the points raised by the experts in the tribunal, including other shipwrights and captains, as well as Hans Clerck, who was responsible for providing rigging for the navy. Clerck noted that the rig was not especially high or wide, and they all agreed that the fundamental problem was that the bottom was too small for the upperworks. It should have been wider, deeper or both.

Several things are clear from the line of questioning and the testimony. Even without the mathematical tools to quantify stability, the maritime professionals present had a very good understanding of the forces at work. They understood the complex balance between hull form, sail plan, weight distribution and gunport location. They knew that warships, with their cannon, had special requirements and they reached essentially the right conclusion about the cause of the accident: there was not enough ship below the water to carry the structure and guns above the water.

It is also clear that most of the participants understood what had happened, and that key people were aware of the problem well before the ship sailed. The line of questioning shows clearly that this was not an investigation of the facts, but like many government inquiries, was political theatre, an attempt to fix the blame. It had been nearly a month since the sinking, and all parties had had time to get their stories straight.

In the event, no one was officially blamed or punished. Captain Söfring Hansson returned to the navy yard as *gårdskapten* and Vice Admiral Erik Jönsson went back to being a master of artillery. Master Jöran Matsson eventually was commissioned and rose to the rank of captain. Hein Jakobsson built three more ships as large or larger than *Vasa* and retired in 1638 with a fine testimonial from the admiralty. Arendt de Groot left the country for a little while, but returned in the 1630s as agent for the great arms merchant Louis De Geer, and became a rich man supplying the Swedish military during the war between Sweden and Denmark in 1643–1645 before moving back to Amsterdam. Henrik Hybertsson's reputation never recovered, and his widow was eventually required to sell off some of their land holdings to pay debts.

In 1629, the Crown cancelled all of its outstanding *arrende* contracts and resumed direct administration of the navy yard. The existing staff were kept on, and Hein Jakobsson became the master shipwright under the administrative

The Dutchman Louis De Geer settled in Sweden and became a leading industrialist, supplying the Swedish military with weapons, ships and other equipment. Arendt de Groot was his primary factor in Sweden for many years.

control of Söfring Hansson. Was the loss of *Vasa* part of the reason that the Crown made this change? The *arrende* system had always had limitations, particularly from an administrative standpoint. Cash flow was difficult to manage, since the entrepreneurs were not required to submit their accounts, and the procurement process was hard to track or measure. With a shift towards annual budgets and proper administrative bureaucracy as part of the growth of Gustav Adolf's fiscal-military state, Crown control may simply have been seen as more efficient. Certainly, the very public loss of *Vasa* was no recommendation for lax administration.

The close connection of Klas Fleming, the architect of the new admiralty, and Söfring Hansson, one of the navy's most able managers, to *Vasa* gave them a unique insight into the problems of shipbuilding. They both probably knew where the blame lay – a modern court would have convicted Captain Söfring of negligence, sailing a ship he knew to be unstable in an unsafe manner, with the gunports open. If he had kept the lower ports closed, *Vasa* probably would not have sunk that day, but she would never have been a good ship. Fleming must also share some of the responsibility. He lacked the political courage to tell the king that his glorious new ship, named for his family, was an accident waiting to happen.

Vasa's dead recovered in 1961 were buried under one of the ship's anchors in the naval cemetery at Galärvarvet in 1963. They were exhumed in 1989 to be studied in more detail, and are now kept in the collections of the *Vasa* Museum.

Consequences

1629

The loss of *Vasa* was certainly a tragedy for the families of the dead and an embarrassment for the Swedish king and his navy. It was welcome news to the king of Poland and the German emperor, and probably not unwelcome to the king of Denmark, even if he was temporarily allied with Sweden. But what actual impact did the sinking have?

In the balance of naval and political power in the Baltic in 1628, one ship, even one as powerful as *Vasa*, was probably not decisive. Sweden lost 18 warships and hundreds of cannon to storms, battle and accidents between the summer of 1625 and the autumn of 1628, more than half of her normal effective naval strength in the 1620s. The Crown only launched four new warships in that period, one of which was one of the losses (*Vasa*). Yet there is no indication that Swedish control of the important lines of transport and communication was ever seriously challenged. Despite a minor victory at Oliwa in 1627, the tiny Polish navy was never a credible threat; the Poles had no way to take advantage of the loss. The Imperial fleet at Wismar never amounted to much, even after it was combined with the Polish fleet. Both potential adversaries were effectively bottled up in their home ports for the few years they were in existence. Denmark had a substantial navy and presented a more formidable opponent, but her resources were depleted by recent defeats in the wars in Germany. The loss of *Vasa* was insufficient incentive to embark on a new adventure.

The loss may have dented Gustav Adolf's pride, but his campaign in Poland was already grinding to a stalemate. It is difficult to see how one ship could have improved his chances of victory, when he already enjoyed effective control of the sea. As a military leader he commanded a certain amount of respect from his friends and enemies, but was still an unknown quantity on the broader stage of European affairs. On the battlefield, he had demonstrated considerable bravery but did not yet enjoy the aura of near-invincibility he would create in Germany after the battle of Breitenfeld in 1631. The loss of *Vasa* occurred before he or the

Gustav Adolf was personally interested in large, heavily armed warships and placed orders for a significant number, despite a lack of effective tactics and against the wishes of his naval high command.

As the struggle between Denmark and Sweden heated up during the 17th century, naval warfare became increasingly important. The battle off the German island of Fehmarn in 1644, a decisive Swedish victory, was only one of many battles between Danish and Swedish fleets, with one side or the other often assisted by the Dutch.

143

The Swedish navy carried Gustav Adolf and his army to the island of Usedom, off the German coast, in the summer of 1630, to intervene decisively in the Thirty Years War. In this painting by Johan Hammer, the king offers a prayer on arriving, in the company of his commanders Gustaf Horn, Johan Banér and Lennart Torstensson.

Swedish military had a significant reputation to lose, and so was probably little more than an occasion for jokes at the expense of the Swedes. A certain amount of satisfaction comes through in the report sent to Christian IV of Denmark just after the sinking by his agent Krabbe, for example.

In terms of strategic and political effect, the timing of the sinking was fortunate for Sweden. It would have been much worse if the ship had sunk two years later, after the country was committed to a large-scale intervention in the main theatre of the Thirty Years War. By then, the navy was trying to protect and supply a bridgehead on the German coast and the king needed all of his prestige and persuasive ability to recruit the Protestant princes of Brandenburg, Saxony and neighbouring states to his cause.

The financial cost of the sinking, the loss of an investment of well over 100,000 dalers, cannot be ignored, but this was a relatively small sum in the context of the coming war in Germany. It was spread over four years and amounted to only a fraction of the cost of hiring a single regiment of heavy cavalry for a year. The loss of time and effort was probably felt more keenly, at least by the shipwrights and other craftsmen who had built and outfitted the ship. Some of the ship's rigging could be recovered and reused, so it was not a total loss, but the gunfounders who had sweated over their furnaces for two years, desperately trying to fill the order for bronze guns, probably had some choice words for the navy.

To make up the numbers in the short term, the Crown purchased or hired ships from abroad, and in 1629, chartered the Ship Company. This was an association of Swedish maritime towns which agreed to purchase or hire ships for the transport of the Crown's goods, and some of the ships were hired directly by the navy for shorter periods. The ships acquired in this way were armed merchantmen, useful as transports or for cruising against merchant shipping, but they were of limited use against warships. Those had to be fought by purpose built warships, built in Swedish shipyards, and the king began placing orders.

The *arrende* system of contracting military procurement to private entrepreneurs had not been completely successful, partly because of the problems of maintaining cash flow through the system and partly because it made government oversight difficult. Late in 1628, the Crown cancelled the outstanding contracts in the shipyards and reverted to direct control of the Stockholm navy yard. One thing that the loss of *Vasa* had definitely not done was to dampen the king's

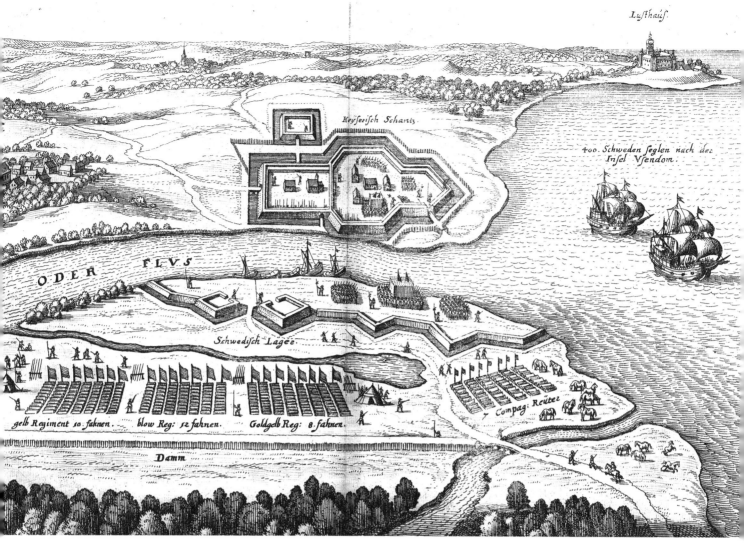

enthusiasm for large, heavily armed warships. As a prototype, it failed, as many prototypes do, but the class of ships which followed served Sweden for a generation.

THE *VASA* CLASS?

When *Vasa* sank, her sister ship had already been launched and was being completed at the navy yard. *Äpplet* (the Apple) had been laid down the year after *Vasa*, and followed her into service in 1629. This ship replaced the failed *Äpplet* built in Västervik and sold back to the builders earlier in the 1620s. Hein Jakobsson had been responsible for the new *Äpplet*'s design and construction from the beginning and made changes from the specification even before he could have known that there were problems with *Vasa*. *Äpplet* remained in service for nearly 30 years, long enough to go through one major rebuilding and to be considered for a second.

The fleet's primary task during the Thirty Years War was to maintain communications between Sweden and the theatre of war in Germany. Here, the Swedish bridgehead at Peenemünde, at the mouth of the Oder River, is protected by Swedish warships.

CONSEQUENCES 145

The approaches to Stockholm, the country's main naval base, were guarded by the fortress of Vaxholm, which was dominated by the tower built in Gustav I's time. Only once during the century, during the Kalmar war in 1612, did Danish ships managed to come this close to Stockholm, but no closer.

Such rebuildings were a common part of a wooden warship's career – most navies planned to make major repairs to a new ship within ten years, due to rot, wear or attack by shipworm. They expected to embark on a major rebuild, which could include significant alteration of the original structure, after fifteen to twenty years of service. With proper maintenance, a well-built ship could last for thirty or more years, and if it was an especially seaworthy or useful ship, it might be worth further rebuilding and preservation.

The surviving survey reports of *Äpplet*'s hull show that she was generally similar to *Vasa* in construction, had the same length and depth, and was initially equipped according to the final armament plan proposed for *Vasa* in the spring of 1628: 56 light 24-pounder bronze guns plus 16 smaller guns on the upper deck. The most significant difference was in the beam. *Äpplet* was 5 feet (1.49 metres) wider than the 17 *aln* (10.1 metres) the contract specified. Hein Jakobsson clearly thought the contracted beam was too narrow, since he had tried to widen *Vasa*. He testified at the inquest after the sinking that he could only add 1 foot 5 inches (0.42 metres) to her beam. *Äpplet* was thus 3 feet 7 inches (1.07 metres) wider than *Vasa*.

An extra metre of beam on a similar hull shape with the same draught should have given *Äpplet* about a hundred tons more displacement than *Vasa*. This offered more space for ballast and gave the ship more "bottom" in relation to the upperworks, as the experts said in the inquest after *Vasa*'s loss. The difference was enough, although after her first period of service, the upper gundeck 24-pounders were replaced by 12-pounders, guns weighing about 25 percent less on average. The weight saved could be transferred to the hold, improving stability further.

After a long career, *Äpplet* was laid up in ordinary. She was surveyed in the late 1650s to determine if she could be rebuilt a second time. Sweden was at war, under intense naval pressure from a Danish-Dutch alliance, and so there was a demand for ships, even older vessels in poor condition. *Äpplet* was badly decayed throughout and not worth saving. She was decommissioned, stripped of useful equipment, and towed out to Oxdjupet. This narrow passage was a choke point on the main route in and out of Stockholm, near the fortress of Vaxholm, about 12 nautical miles from the city centre. The Crown wished to close this passage so that enemy ships approaching the capital would have to pass within range of

the guns of Vaxholm. *Äpplet* and a pair of captured Danish ships from the war of 1643–1645 were deliberately sunk in Oxdjupet, the first of many old warships deposited there. She is probably there still.

Gustav Adolf had a distinct preference for larger ships, while his council and many of his admirals preferred smaller vessels, both for financial and tactical reasons. Until the end of his reign, he continued to order ships of increasing size to be built in Stockholm, Västervik and Göteborg. These ships were not especially well suited to the tactical environment of the Baltic, and it has been thought that the king may have been contemplating challenging the naval powers of the Atlantic, especially Spain. Many of the later contracts were cancelled after the king's death in 1632, but Hein Jakobsson completed two which entered service in the 1630s. *Kronan* was laid down in 1631 and launched the following year. At over 1500 tons displacement, she was the largest ship in the navy until she was replaced by a new ship of the same name in the 1670s. *Scepter*, laid down in 1632 and launched in 1634, was slightly larger than *Vasa* but of different proportions. Master Hein's three ships were named for the traditional symbols of the king's power, the orb, scepter and crown, and belonged to an administrative class of warships called *regalskepp* or *realskepp*, the largest ships in the navy. Had *Vasa*

A ship cemetery lies on the bottom of Oxdjupet, off the island of Rindö near Vaxholm. *Vasa*'s sistership *Äpplet* was sunk here with many other ships to block the passage and force ships to sail past Vaxholm. The shore is studded with fortifications from many eras.

CONSEQUENCES 147

The battle off Kolberger Heide on 1 July 1644 ended in a draw, but became famous in Denmark for King Christian IV's participation. He lost an eye on board *Trefoldigheden* but called on his men to fight on. Wilhelm Marstrand painted this heroic image in 1864, immediately after the Prussian defeat of Denmark in the Second Schleswig War.

survived, she would have been in the same group, but the term was never applied to her while she was afloat. After *Scepter*, relatively few ships were built during Chancellor Axel Oxenstierna's regency (until the young Queen Christina came of age in 1644), but efforts were made to make sure that the navy was properly manned and fed through reform of the recruitment and provisioning systems.

Up to 1644, naval operations were much as they had been in the 1620s. Gustav Adolf had shifted his focus from Poland to Germany, landing near Peenemünde in the summer of 1630 with an army. A large fleet transported the king and his forces and remained active on the German coast, supplying the bridgehead and maintaining open sea lanes between Sweden and the battle front. *Äpplet* was not part of this fleet, but *Vasa*'s former captain, Söfring Hansson, commanded one of the ships. There was no significant challenge on the water, and from 1631 the campaign moved farther south, into central Germany, where new troops and supplies were raised locally rather than carried over the sea. In 1632, the Imperial port of Wismar was captured, and with it the remains of the Imperial and Polish fleets, helping to make good some of the losses of the 1620s. This included the recapture of the former Swedish ship *Tigern*, lost to the Poles at the Battle of Oliwa in 1627.

Even after Gustav Adolf's death at Lützen, Swedish armies continued to be a significant and eventually decisive force in northern Europe until the Thirty Years War ended in 1648. Swedish successes after 1635 emboldened Oxenstierna to take up a project he had cherished for many years, the conquest of the Danish provinces on the southern border of Sweden. Halland, Blekinge and Skåne in the southwest were traditionally part of Denmark, and allowed the king of Denmark to control Öresund, the passage into the Baltic. Oxenstierna wanted to drive the Swedish frontier all the way to the water and break Danish control of the passage. In this he had the sympathy of the Dutch merchants who dominated Baltic trade – they wanted to force down the toll rates charged by Denmark.

Oxenstierna planned an amphibious campaign, in which a Swedish army operating in Germany would march north to invade Denmark while the Swedish navy, assisted by hired Dutch ships, would destroy or cripple the Danish fleet. King Christian IV would be forced to make large territorial concessions and the Danish fleet would be eliminated as a potential challenger to Swedish control of the Baltic. The war was brief, lasting only two years, but was notable for the role played by the navy. At last the big ships ordered by Gustav Adolf might see the use for which they were intended.

The central campaign of the war was fought in 1644. The land action was inconclusive (the Swedish army could not take itself away from the German war long enough to be decisive), but two naval battles shifted the balance of power in the Baltic significantly. At the end of June, a Swedish fleet of 34 major warships plus smaller vessels awaited the arrival of a Danish fleet of 25 warships plus transports in the narrow straight between the German coast east of Kiel and the southern islands of Denmark. The Swedish fleet was not only larger, but superior in firepower, although most of the ships involved were relatively old, almost all of them ordered by Gustav Adolf and many contemporaries of *Vasa*. Each fleet was divided into three squadrons: avantgarde, centre and rearguard. On the Swedish

Christian IV's bloody clothes from Kolberger Heide are displayed at his palace of Rosenborg in Copenhagen.

1644

THE BATTLE OF FEHMARN

In the autumn, after the inconclusive battle at Kolberger Heide on July 1st, the Swedish and Danish fleets met again in a much more violent and decisive battle. Smaller fleets were involved, and the Swedes had the assistance of a Dutch squadron of armed merchantmen. Ten Danish ships were captured and two more sunk, an unusually lopsided result.

The Danish ship *Lindormen* is set on fire by the fireship *Meerman*. Many of the crew saved

The Swedish flagship *Smålands Lejon*, built in Västervik in 1633, was damaged and withdrew. The Swedish admiral on

The Swedish ship *Nya*

13 OCTOBER

The Danish flagship *Patientia* is boarded by soldiers from the Swedish

The Danish admiral Pros Mund refused to surrender and fought the boarders

The Swedish admiral Karl Gustav Wrangel, according to tradition, took his speaking trumpet home with him from Fehmarn to his castle Skokloster, north of Stockholm, where it may be seen today. Communication between ships during a battle was often nearly impossible.

side, *Scepter* was the flagship of the avantgarde and *Kronan* led the centre. Vice Admiral Klas Fleming commanded the fleet and hoisted his flag in *Scepter*. The Danes arrived on July 1st, about noon, and found the Swedes to windward, near Kolberger Heide in the eastern approaches to Kiel Bay, the ideal position from which to control a battle. Fleming led four attacks on the Danish fleet, but was unable, despite his superior force and the weather gauge, to draw the Danes into the all-out general action which would allow his firepower to be used to full advantage. Casualties were relatively light, and the fleets withdrew. Although unsuccessful, the action was an early attempt to coordinate a fleet in battle.

Fleming retired west into Kiel Bay, where provisioning became difficult, seamen became sick and the auxiliaries hired from Holland failed to show up. The Danish fleet followed, set up shore batteries and fired on the Swedes, killing Klas Fleming. Under Karl Gustav Wrangel (a general with experience of amphibious operations), the Swedes eventually sailed out of the bay to engage the Danes, but the Danish fleet declined to fight. The Swedish fleet sailed back to the Stockholm archipelago with the expectation that the fighting was done for the year.

In August the Dutch ships finally arrived, but would return to Holland at the end of the autumn. It was decided to make the most of the combined force while the weather lasted. A smaller Swedish force, consisting of the best ships and the best officers, was hand-picked to combine with the Dutch in a sortie south from Kalmar, under Wrangel. He sailed on October 5th, with 14 major warships and 19 armed merchantmen. *Scepter* and *Kronan* were not included, nor were any of the other big ships. They met the Danish fleet, much reduced since the summer, on October 13th near the German island of Fehmarn, farther to the east from Kolberger Heide. The Swedish-Dutch fleet was larger, with greater overall firepower, and a manpower advantage of better than two to one, but the Dutch vessels, armed merchantmen, could not stand up to the firepower of the Danish warships.

The tactics used in the battle were the same for which *Vasa* had been built. Rather than fighting as a fleet, each ship attacked another ship of similar size in a swirling melée, while the Dutch rounded up stragglers. The battle was essentially a series of single-ship actions, in which the Swedish superiority in crew numbers, rather than gunnery, was the decisive factor. The Swedes were more aggressive than at Kolberger Heide and the Danes lost ten ships captured and two others destroyed. Fehmarn is one of the great victories in Swedish naval history, with far-reaching consequences.

The transfer of ten ships from one side to the other changed the balance of naval power in the Baltic. Even though Swedish plans for the invasion and conquest of Denmark never came to fruition, Christian IV was still forced to sue for peace. In the settlement of Brömsebro in 1645, Sweden acquired control of the province of Halland, as well as the strategic Baltic islands of Gotland and Ösel (Saaremaa), losses from which Denmark never really recovered. Two other Danish provinces, Skåne and Blekinge, as well as the Danish-Norwegian Bohuslän, fell to Sweden in 1658.

The fleet, including *Scepter* and *Kronan* but not *Äpplet*, which was by now rotten, was in action again during the next war with Denmark in 1655–1660. This time, the Dutch sided with Denmark to prevent a Swedish blockade of Öresund and sent battle-hardened fleets to the Baltic in 1656 and 1658, which included one of the great sea commanders of the 17th century, Michiel de Ruyter. The Swedish navy acquitted itself reasonably well, especially against the more experienced and more modern Dutch, but it was clear that a new tactical model was needed. The melée tactics of the 1644 battles would not succeed against newer ships fighting together as a formation, concentrating massed firepower against selected targets.

Once this war came to a close, Gustav Adolf's big ships were reaching the end of their practical service lives. *Kronan* had played a significant role in the battle of Öresund in 1658, but they were old-fashioned and starting to decay. Although *Äpplet* was condemned during the war, *Scepter* and *Kronan* lasted another decade, until they were decommissioned in 1671. They followed *Äpplet* into Oxdjupet in 1675. *Äpplet* did not see as much action as the others, and she may not have been a big enough improvement on *Vasa* – she seems to have spent much of her career in the reserve squadron stationed in the Swedish archipelago. The other two were decent sailers, although neither had *Vasa*'s firepower.

They may have been somewhat ahead of their time. Their firepower was certainly an advantage in a gunnery duel, but Wrangel's decision not to include them in the smaller, elite fleet which fought at Fehmarn in 1644 suggests that ships with two gundecks were not as well suited as the more traditional single-deckers to the boarding tactics employed by the Swedish navy. If they had been

The Danish toll at Öresund was a constant irritation for the Dutch merchants who dominated Baltic trade with the West so the Danish-Swedish wars were of great interest to the Dutch. In this Dutch political cartoon, the Danish king Christian IV and the Swedish field marshal Lennart Torstensson play backgammon over the outcome of the war of 1643–1645.

built as part of Gustav Adolf's dream of expanding Swedish naval power beyond the Baltic, it was a dream that died with him in 1632, leaving the navy with a few large, expensive ships which were prestigious but difficult to use operationally.

TOWARDS THE 74-GUN SHIP

Vasa and the other big ships of her era pointed the way towards the future of naval warfare. They did not fit the hand-to-hand philosophy of their time, possibly because they had too little usable deck space for the size of their boarding crews. The sides of warships of this period leaned inward from the waterline, so a higher ship had a narrower upper deck for the same maximum beam. Still, the emphasis on firepower and gunnery which multiple gundecks implied was the way forward. In many ways they were a new weapon in search of appropriate tactics. Those tactics began to emerge in the North Sea in the middle of the 17th century, developed intensely in a series of naval wars between England and Holland in 1652–1674. Gunports were redesigned so that all of the guns on one side of the ship could be trained on the target simultaneously and fired as a devastating broadside. Captains learned to communicate and cooperate with each other in battle. The primary tactical model was referred to as the line of battle, ships sailing in a continuous line, so that they could provide supporting fire to each other.

With the shift to line tactics for fleet actions, the search was on for the ideal combination of firepower, speed and seaworthiness. Enormous ships with three full gundecks and over 100 cannon were built, and while every major navy had a few, they could be unwieldy and the lowest tier of gunports was often so close to the water that is was necessary to keep the ports closed in anything but fine weather. Ships with only a single gundeck did not have enough firepower to be competitive in the line, but were still necessary for cruising, independent action to suppress piracy and smuggling or to hunt the enemy's merchant shipping. Two gundecks became the standard for the ships which formed the core of the line, as such ships could be fast and seaworthy but still carry enough guns to be effective in the line. They were also substantially cheaper to build and operate than three-deckers, due to their smaller crews.

The number and type of guns also had to be settled. A two-decker (which meant two gundecks, but such ships usually had a few guns on the upper decks

The pinnacle of development of the sailing line-of-battle ship was reached in the later 18th century with ships following *Vasa*'s basic configuration of two gundecks plus an armed upper deck. This is the British ship *Agamemnon*, built in 1781, Admiral Nelson's favorite ship. She was a Third Rate mounting 64 guns, present at the battles of Copenhagen (1801) and Trafalgar (1805).

as well) could carry anything from fewer than 50 guns to more than 80, but the distance between the gunports was more or less fixed by the ergonomics of loading and firing the guns and the structural needs of carrying them. Thus more guns meant a longer ship, and there were practical limits here as well. Through a long process of experiment, the navies of England and France eventually settled on 74 guns as the ideal compromise by about the middle of the 18th century, and other navies followed suit.

Gustav Adolf's idea of unitary armament, two full decks of identical guns, turned out to be impractical for reasons of weight distribution, and *Äpplet* only sailed in this configuration for a short while. Unitary armament within one deck was adopted, and by the end of the 17th century, most larger navies had arrived at a system of related guns in just a few sizes, similar to Gustav Adolf's standardization of his land artillery on just three sizes of gun. This was complicated slightly by the normal consequences of naval battles, the capture and use of enemy ships and guns, since every nation had its own, unique standard, but the overall concept was similar. The heaviest guns were carried on the lowest deck, with progressively lighter guns on the decks above, much as *Äpplet* had been re-armed.

Where later ships differed was in size. At displacements of less than 1500 tons, ships like *Vasa* and *Äpplet* were simply too small to carry so much firepower. *Scepter* and *Kronan* carried fewer guns (58 and 68 guns, respectively, at Kolberger Heide in 1644) than planned for *Vasa* and *Äpplet* (72) even though *Kronan* was a markedly larger ship. By the later 17th century, ships carrying so many heavy guns were much larger, with displacements over 2000 tons. They were not much longer than *Vasa*, but wider, deeper in the water and lower above the water. This was partly related to the ascendancy of English and Scottish ship designers over the Dutch in the later part of the century, but also a general trend in naval shipbuilding throughout northern Europe. *Vasa* was a harbinger of what was to come, but built without any experience of how two-deckers would behave. Her loss was a tragic but necessary element in building up the knowledge needed for the development of the ship of the line, the pinnacle of naval technology for nearly two centuries.

The Swedish navy eventually became oriented towards coastal defence and inshore waters. The new class of 60-gun ships of the line authorized in 1780 were derived from F.H. Chapman's experimental liner *Wasa*, the first ship to be named thus since the loss of *Vasa* in 1628. They were small but heavily armed, with 36-pounders on the lower deck and 24-pounders above.

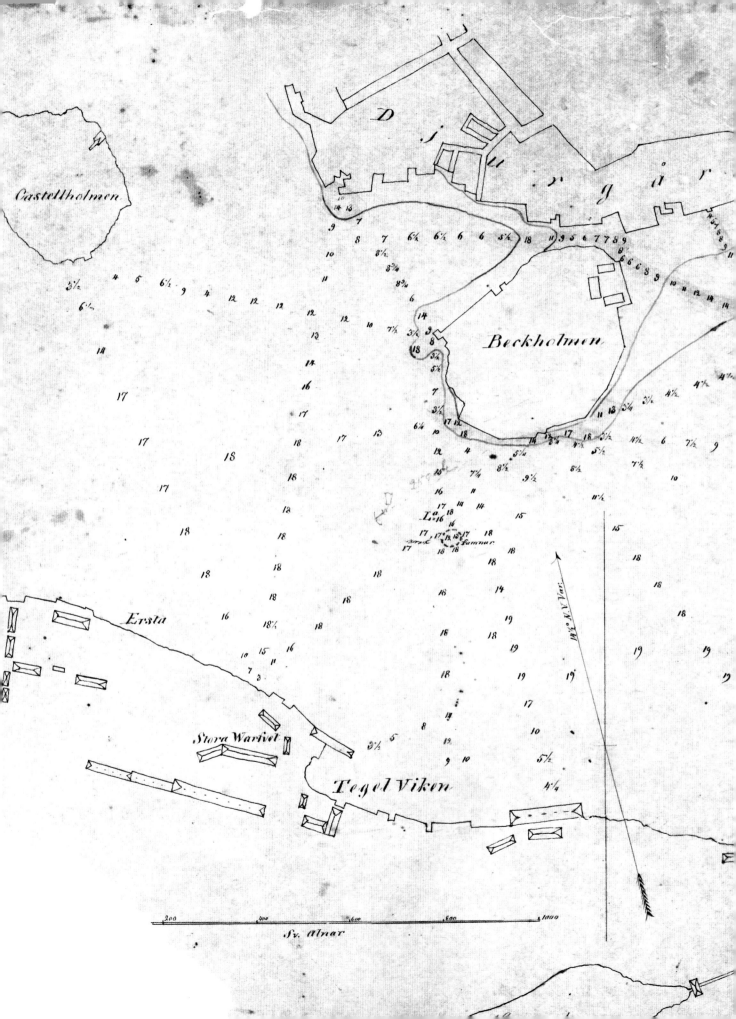

Lost but not forgotten

1664

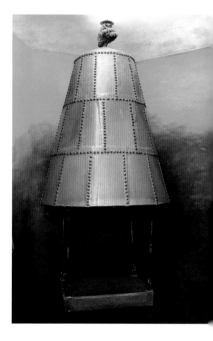

Reconstruction of a 17th-century diving bell in the *Vasa* Museum. It is built of wood sheathed in lead, with a heavy lead block hung under it for weight. The diver stood on the block or sat on a saddle mounted inside the bell.

Hand-drawn copy of the Stockholm harbour chart of 1836 with the text "wreck" written next to a bump in the bottom off Beckholmen, exactly where *Vasa* lay.

It was the cold. There was enough air in the diving bell for a half hour, but no one could take the cold for more than about a quarter of an hour. He was named Juhani, but because he came from Finland, everyone called him Finn. He was a hard man, used to bitter winter temperatures in a harsh northern climate, but this job took a special kind of toughness. Getting dressed in the woollen undersuit and the oiled leather oversuit, clamped with iron bands at the thighs and waist, made most of the divers break out in a sweat during the summer months, but it did not last long. When the bell was put over the side of the diving barge, the water started at mid-thigh and no amount of greasing would keep all of it out. It seeped in at the seams, and soaked the wool underneath.

As he was lowered into the dark, the rising pressure forced the water ever upwards in the bell, until it was at chest level when he had reached the deck of the sunken ship. At the bottom, the water was just above freezing year round and the pressure pushed it in through the stitch holes like hundreds of needles. Despite the heavy exertion of salvage, he rapidly lost the feeling in his feet and hands. It was now the second season of work, and he knew how long he could last, when it was time to tug on the signal line so the crew on the dive barge would start to raise the bell. He would get colder on the long journey back to the surface, sitting still on the saddle mounted in the bell, and when he could finally climb out from under the bell and into the sunlight, he was often shivering. As soon as he could peel off the sodden diving suit and get into dry clothes, he would have a pipe and a little *brännvin* (spirits) to try to get some heat back into his body, preparing himself for the next dive.

The work at the bottom did not go quickly. Aside from the cold, it was nearly pitch dark and the water was muddy, so everything had to be done by touch. Usually, the diver could stay in the bell and work with long-handled tools, but for some tasks he had to hold his breath and crawl out of the bell. It was very important to keep one hand on the bell or have a line tied to the saddle,

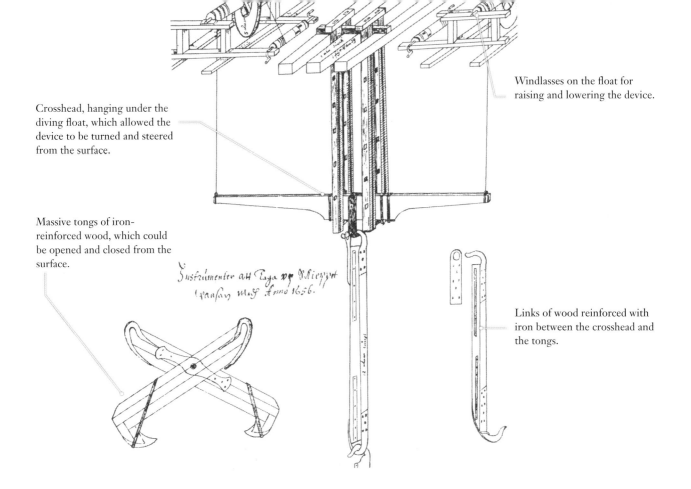

Crosshead, hanging under the diving float, which allowed the device to be turned and steered from the surface.

Windlasses on the float for raising and lowering the device.

Massive tongs of iron-reinforced wood, which could be opened and closed from the surface.

Links of wood reinforced with iron between the crosshead and the tongs.

Lifting device constructed by Andreas Peckell for recovering *Vasa*'s guns. The drawing is from 1665 (misdated 1656).

otherwise it would be nearly impossible to find the way back.

It had taken all of the first season just to clear the cable, chain and debris off the wreck. It seemed that the ship was covered in anchors and miles of rope, and in the early stages there was a constant danger of getting tangled. They had managed to haul up most of the guns from the upper deck this year, although it had taken some experimentation to find a reliable way to do this. They had initially raised the guns with their carriages, but the divers soon discovered it was both faster and safer to break open the corroded capsquares, the iron straps holding the tube to the carriage, and leave the carriage behind.

With the upper deck cleared, they had torn up the planking to get to the main prize, the ship's heavy armament in the gundecks. The engineer, Andreas Peckell, had developed special tools for this job, and they had taken up 30 cartloads of timber. Once they broke through, it had been relatively easy to get the big guns out of the upper gundeck. Peckell had made another special tool, an enormous set of tongs which could be opened and closed from the barge, for lifting the guns. All the diver had to do was guide them into place and make sure they closed securely around the gun, then get out of the way.

Peckell had also figured out a way to get the cannon out of the lower gundeck. The upper gundeck was too solid to break through, so they would pull the cannon out through the gunports. This was going to be harder now that Peckell was gone. He was an immensely clever man, with a stream of ideas for machines and tools, but he did not get along very well with Albrecht von Treileben, the leader of the investors backing the project and the man who held the salvage privilege from King Carl XI's regency government. There had been a big fight on the dive barge in the middle of the summer, and the Göteborg divers loyal to von Treileben had forced Peckell to leave, but kept all of the tools. Von Treileben was now running the project.

It was late in the autumn and the weather was deteriorating. They would have to stop diving soon and come back in the spring. Juhani would go home to Finland, to live through the much more bearable cold of the long, snowy winter. At least he would take home plenty of money, which made the freezing needles of muddy water almost tolerable.

THE PRIZE ON THE HARBOUR BOTTOM

As soon as *Vasa* sank, attempts began to recover the ship. The Crown had invested a considerable sum of money in the ship and its armament, and neither the hull nor the guns would be significantly damaged by days, weeks or even months under water. There was no complex machinery to rust and seize up, and it was possible to put a sunken ship back into service relatively quickly. The window of opportunity was not without limits, as the iron nails and bolts holding the ship together would start to corrode relatively soon after a ship sank. Eventually, once the fasteners were weakened, it was not worth raising the hull, but it could be a matter of years in the right conditions. The bronze guns could survive decades under water, and even after they had corroded beyond a usable state, they still had substantial scrap value, worth the investment in a recovery operation.

The technology for raising ships was well understood by the early 17th century. A sunken wooden ship actually weighs very little, as the buoyancy of the water acting on the wood of the hull cancels much of the weight. Out of the water, *Vasa* weighed just over 1200 tons with all of its ballast, guns and equipment, but once submerged, the weight was closer to 200 tons, so the lifting force needed was relatively small as long as the ship had not embedded itself in the bottom.

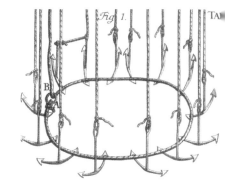

The normal method for raising sunken ships, in use since the Middle Ages, was to connect the ship to lifting pontoons at the surface with anchors, sometimes with a rope girdle as here. When *Vasa* was raised in 1961, there were many anchors still fastened in the structure, remains of the 17th-century salvage work.

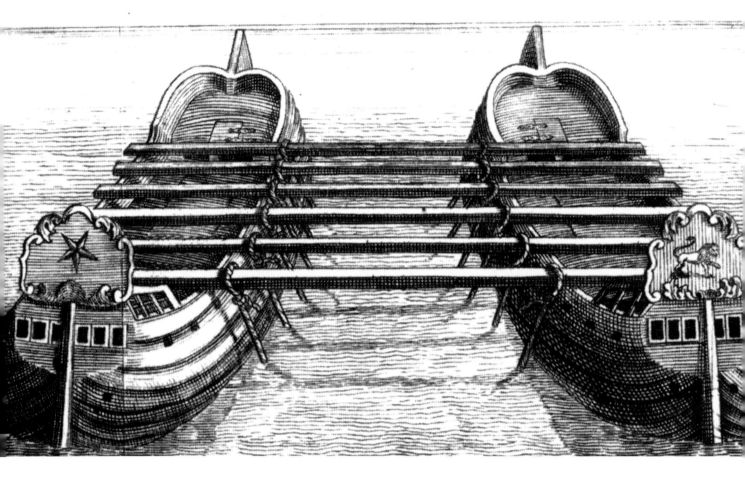

Ships, stripped of excess weight, were used as lifting pontoons. The pontoons were filled with water to sink them partly, the cables to the wreck were tightened, and the pontoons were pumped out, lifting the wreck.

The normal practice was to fasten a large number of anchors into the sunken hull and take their cables to two ships moored over the wreck. The two floating ships were pumped full of water, until they were close to sinking, and the cables were tightened. The water was then pumped out of the floating ships, slowly but steadily increasing their buoyancy and eventually lifting the sunken ship off the bottom. Once free, the ship was moved into shallower water and grounded again, where the process was repeated. Through a series of such lifts, the sunken ship could be brought to the surface, where it could be pumped out until it floated on its own keel.

If the ship did become embedded, the recovery was much more difficult. The suction between the hull surface and the glacial clay of the sea bottom greatly exceeded the submerged weight, so much more lifting force was needed. If the suction was too great, the lifting cables would fail or the anchors might tear out of the hull.

The Royal Council granted a salvage privilege to Ian Bulmer, an Englishman, on the 13th of August 1628, just three days after the ship sank, in hopes that the ship could be raised before the king returned from Poland in the autumn. Bulmer succeeded in righting the ship from its list to port, the necessary first step before the hull could be raised, and he had the use of the largest warships in the harbour as lifting barges, but he could not get the ship up. He was replaced by a well-known Dutch engineer, Willem de Besche, who had carried out many complicated technical projects, but he did not succeed either. When winter came, *Vasa* still lay at the bottom. Her topmasts had been taken down at some point in the autumn, else the winter ice would break them, so there was nothing for the king to see when he came home in December.

Preparations were made in the winter for a major effort in the spring of 1629. The resources of the navy yard were committed to the project, and the smithy there spent the winter forging more anchors to be used in the recovery attempt. Captain Söfring Hansson had been assigned by the king to coordinate and facilitate the salvage almost immediately after the sinking, possibly assisted by *Vasa*'s master, Jöran Matsson.

Throughout the spring and summer, repeated efforts to get the ship up failed. The reports from the salvage operation frequently mention broken cables and equipment. This suggests that the lifting force at the surface could not overcome the weight and suction at the bottom, but did exceed the strength of the hemp and wrought iron of the lifting gear. Bulmer's initial righting of the ship had probably contributed to the problem by pushing the hull farther down into the organic mud and glacial clay of the harbour bottom. Hope began to fade as the summer wore on, and the effort was eventually abandoned.

In the years that followed, well into the 1630s, one entrepreneur after another approached the Crown with a plan for raising the ship, and a number of privileges were issued, but nothing practical ever materialized. By about 1640, it was realized that the hull was probably so decayed that it could not be put back into service if it was raised and people began to accept that the ship was lost. The guns, however, were a different matter.

In 1652, Colonel Alexander Forbes, a Scottish army officer in Swedish service, applied to Queen Christina for the salvage rights to sunken ships in Swedish waters, which he was granted for 12 years. The following year he received a specific privilege to salvage material from *Vasa*. He employed several engineers, including Ian Bulmer, but failed to accomplish anything of substance.

In 1658, Hans Albrecht von Treileben, an army officer, successfully recovered guns from the Danish ship *Sophia*, sunk off Göteborg. He used a new invention,

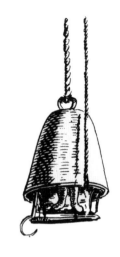

The Italian priest Francesco Negri was an eyewitness to the diving work on *Vasa* in the 1660s and described the use of the diving bell in his account of his travels, *Viaggio settentrionale*, even providing an illustration. The long-handled hook was a standard diver's tool for recovering objects and manipulating rope.

Lifting claws, similar to Peckell's, illustrated in Mårten Triewald's 1734 treatise on diving and salvage work.

Of *Vasa*'s 64 cannon, 61 were recovered in the 17th century. The remaining three were found during the salvage and excavation in 1956–1961 and are now exhibited in the *Vasa* Museum. This 24-pounder lay on the remains of its carriage in the lower gundeck.

the diving bell, to do this, and demonstrated the device to King Karl X Gustav. Free divers, holding their breath, could work in depths in excess of 20 metres, but their useful time on the bottom was measured in seconds. By the Renaissance, scientists and engineers were looking for ways to extend the useful bottom time. The first practical technology developed was the diving bell, a metal vessel large enough for a diver, open at the bottom with heavy weights chained to the lower edge. When lowered into the water, it trapped a bubble of air inside, which the diver could breathe. A diver could work from inside the bell for a considerable period, depending on the depth.

The king granted von Treileben and his associates the rights to sunken ships on the west coast and elsewhere in Sweden, but specifically excluded *Vasa*, since Forbes had those rights. Through his technical partner, a Scottish diver named Jacob Maule, he came to know the German engineer, Andreas Peckell. Peckell had his own inventions for underwater work, and had successfully cleared the harbour of Landskrona of ships sunk there by the Danes to block it.

Von Treileben had been making use of his existing grant to recover guns, equipment and coins from several sunken ships, and he had powerful friends at court. By the summer of 1663, he had managed to have Forbes's privilege cancelled and the rights to *Vasa* were his. He took Peckell into partnership that summer (Peckell would receive a quarter of the value of everything recovered), and in the autumn, they embarked on a project to recover as much as they could from the ship. They claimed that they would raise the ship as well, but this may have been more bravado than realism.

They employed a dozen men at first, many of them divers from the west coast of Sweden whom von Treileben had used on his earlier projects. Peckell taught them the new methods he had developed, including his technique for pulling the guns out through the gunports with a giant set of tongs, operated from the surface. These were suspended from the dive barge and would be lowered to the bottom, where a diver would place them around the muzzle of a cannon projecting from the side of the ship. The tongs were closed by cables operated from the surface, gripping the end of the cannon. Using windlasses at the surface, the tongs could be lifted to pull the gun up out of the brackets on its carriage and then manoeuvred away from the ship to pull the gun out. The process seems to have been relatively smooth, as there is only minor damage to the gunport edges, despite the narrow clearance between the port and the gun.

The initial work of clearing all of the old salvage debris from the ship took up the autumn of 1663, and they could not begin to raise any guns until the following spring, with the first gun coming up on April 1st, 1664. Once his divers were familiar with the engineer's techniques, von Treileben was eager to be rid of his partner and had his divers forcibly remove Peckell at the end of July. Peckell's barge and tools were confiscated, and he was left with little more than his clothing. Von Treileben had recovered all of the guns considered accessible by the late summer of 1665. Some could not be reached, and one was dropped onto the upper gundeck and could not be recovered, but the majority of the guns were brought up.

Peckell was understandably angry, and took his former partner to court for the money he was owed. Thanks to this case, which dragged on into 1666 and required the testimony of all of the participants in the project, a detailed record of the recovery operation survives. Peckell sued for his quarter share of everything

Despite the rough nature of the work and the crude methods of the 17th century, it was possible to pull the cannon out through the ports of the lower gundeck without significant damage to the surrounding structure. This bite out of a gunport sill is one of few injuries done in the process. In contrast, most of the upper deck was torn up and brought to the surface in order to get the guns out of the upper gundeck.

Surfaces exposed to the current were heavily eroded, as on the left side of this image. Other areas were protected by other objects or the vagaries of the current, as on the right.

that was raised before he was removed. Von Treileben, who must have known he was on shaky legal ground but had experience of getting a favourable verdict through his political connections, lobbied his friends on the Royal Council to intervene in his favour. This time he had tried to pull one string too many; not only did Peckell win, but the court ruled that since von Treileben had kept (effectively stolen) Peckell's tools and used them to raise the rest of the guns, Peckell was entitled to a quarter of everything recovered over the duration of the project. This probably consumed most of von Treileben's profit.

What then became of the guns? The trial records do not agree on how many guns were recovered, but the total was probably 60. Customs records from 1665 show that von Treileben's financial backers exported a total of 53 bronze cannon from *Vasa* to Lübeck in the summer and autumn of that year. There was a lively market for guns in Hamburg and Lübeck, as the Danish king had sent his agents to northern Germany in search of cannon for his expanding navy. It is thus possible that *Vasa*'s 24-pounders were eventually fired in anger, but at the Swedish fleet in the Scanian War of the 1670s, when Denmark attempted to recover the provinces of Skåne, Halland and Blekinge, lost to Sweden in the 1640s and 1650s.

One more gun was recovered from *Vasa* in 1683, by Jörgen Liberton, before diving work on the ship ceased. All told, 61 of the ship's 64 bronze guns had been recovered. Peckell and von Treileben were among the first to demonstrate that the diving bell was a commercially viable technology. The bell was used to salvage guns from other wrecks, and von Treileben's initiative led to the establishment of the first permanent, professional diving salvage companies in northern Europe.

THE LONG SLEEP

The attempted salvage of the ship and the recovery of the guns caused significant damage to the ship as it lay on the bottom. The mainmast was broken off, the upper deck torn up, the upper works damaged, and the material left in the upper gundeck stirred around. Gun carriages from the upper deck were removed with their guns or thrown down into the wreckage of the upper gundeck. But while Bulmer, Peckell and von Treileben had been doing their worst, slower natural processes were combining both to break down and to preserve the ship.

In marine environments, the worst enemy of wooden shipwrecks is a boring mollusc called *Teredo navalis*, the shipworm. Along with other boring organisms, it chews through waterlogged wood, tunneling deep in large colonies until there is no wood left. In the tropics, a ship can disappear except for a mound of stone ballast in just a few years. *Teredo* cannot live in the cold, brackish waters found in most of the Baltic Sea, so *Vasa* escaped this fate. Erosion bacteria attack wood cells at a microscopic level, but they require oxygen to live. The water at the bottom of Stockholm harbour is nearly anaerobic, without oxygen, and until the 20th century was heavily polluted by raw sewage dumped directly into the harbour from Stockholm's thousands of privies. Erosion bacteria may survive in this environment, but their activity is drastically reduced. Natural chemical processes that occur as the wood becomes increasingly waterlogged cause cellulose, the main strength component of wood fibres, to break down and thus lose strength, but the process is very slow. The pollution in the water helped to protect the wood while it was submerged, but chemicals built up in the wood that would cause trouble once the ship returned to the air.

When *Vasa* was raised, the ship was full of mud and debris deposited over the centuries, which protected many of the more delicate objects inside the ship.

All of this meant that the wooden parts of the hull, together with thousands of wooden objects, survived in outstanding condition, deteriorating very slowly. The outer few millimetres of wood surface decayed, but the wood beneath in the main structural timbers retained much of its original strength. Recent research has shown that there is also significant decay in the innermost core of many timbers, but their great size means that there is still a significant amount of material with some remaining strength. The heavy construction, while it contributed to the ship's lack of stability and ultimately its loss, was responsible for helping to preserve most of the ship as an intact structure. It maintained the shape of the structure over the centuries that it lay on the bottom, and it discouraged von Treileben's divers from trying to break up more than the light pine planking of the upper deck.

The water in which *Vasa* sank is in reality an outlet which drains the great lake of Mälaren and the runoff from the surrounding countryside into the Baltic, so there is a current over the site. This water carries a great deal of sediment with it, and this sediment, flowing over exposed wood, acted like a low-speed sand blaster, grinding away the surface. Areas affected by erosion bacteria and the dissolving of cellulose were softer and more susceptible to abrasion, so as they decayed they were worn away by the current. Areas protected from water flow, such as parts of the ship buried in the bottom or surfaces covered by other objects, did not suffer as much. The ship itself helped to preserve the wood, because it disturbed the current flowing over it, causing the sediment to fall out of the water and cover the artefacts lying on the decks inside the ship. Once objects were buried, the decay process slowed down even more, and this is why delicate objects, such as woollen clothing and hemp sails, have survived.

Unfortunately, the conditions which are so beneficial for organic materials are especially harsh for metals, especially iron. All of the metal in the ship is corroded to some degree, but virtually all of the wrought iron disappeared completely. Of nearly 8,000 bolts used to hold the ship together, only a handful survived. More than 250 woodworking tools were found in the ship, but all that survives is the wooden parts: many axe handles, no axe heads. The anchors, both the ship's original four and about forty others used in the attempted salvage, survived due to their larger mass. Cast iron also survived, due to the high concentrations of carbon, but what remains is mostly just the carbon. There are over 600 24-pounder cannonballs, but most now weigh less than half their original weight. Copper and lead objects fared better, since they are more resistant to corrosion, but tin and silver did not. The few silver coins found in the ship are mostly unidentifiable, ragged black discs. The metal corrosion products, especially rust from the iron cannonballs and bolts, permeated the organic materials, and now remain in the wood as a catalyst for the formation of acids, which will attack the wood if not neutralized.

On the other hand, the swift corrosion of bolts and nails contributed to the outstanding preservation of some of the most spectacular wooden objects, the sculptures at the bow and stern. These ornaments were attached to the hull almost exclusively by light nails, which corroded through very quickly and allowed the sculptures to fall down into the mud. The mud protected them from erosion or other degradation, and so many survived with crisp toolmarks and even paint. The diffusion of iron corrosion products into the surrounding wood also made the affected areas very hard and more resistant to erosion by the sediment-laden current.

It is clear that once the parts held on by nails or rope had fallen off – this included the blocks and deadeyes of the rigging, the after face of the transom, the gunport lids and all of the beakhead – the rest of the structure was quite stable. Once Liberton left the ship in 1683, it seems to have remained undisturbed for

Iron fared much worse than other materials at the wreck site. Only large objects such as these anchors survived, and even they were much reduced by corrosion.

over 150 years, embedded up to the waterline in the bottom mud with much of the sterncastle and most of the foremast still standing. Occasionally a ship would drop an anchor into the wreck, and some of the everyday trash dumped into the harbour by the residents of Stockholm found its way into the gundecks, but otherwise *Vasa* slept through the years, slowly eroding, but not forgotten by any means.

RENEWED INTEREST

In the 19th century, *Vasa* began to be visited again. Ever larger ships came into the port of Stockholm, and if they dropped an anchor into the wreck, they had the power to pull it free. At some point, one such anchor was caught in the sterncastle and torn out again, breaking off the upper part of the stern and dragging it away to port. Larger ships required new quays for loading and unloading and large drydocks for maintenance. Because Stockholm is built on bedrock, this meant blasting. Some of the resulting rubble was dumped in front of Beckholmen, landing on *Vasa*. This completed the destruction of the sterncastle.

Diving technology was relatively static through the 18th century, but at the beginning of the 19th century, advances were made in the development of diving dress, a waterproof suit with a helmet to which air was pumped from the surface. A diver could now stay on the bottom for a much longer time and move around freely. This opened a new era in underwater construction and salvage, soon exploited by navies and private salvage firms. The new diving dress was introduced in Sweden in the 1840s by a naval officer, Anton Ludwig Fahnehjelm, and one of the ways he demonstrated its effectiveness was by diving on the wreck of *Vasa*.

How had he known where it was? Even though it had not been the scene of active diving in over a century, the story of *Vasa*'s ill-fated maiden voyage had passed into local lore and was regularly used as an object lesson for new naval officers. The first history of the Swedish navy, written in 1734 by Carl Bechstadius, mentions the wreck, as do most subsequent naval histories of Sweden. The official chart of Stockholm harbour from 1836 has the location of the wreck marked, although not identified by name. After Fahnehjelm, divers and salvage firms visited the wreck sporadically for the rest of the 19th century. A major diving operation was carried out there in 1895–1896, and once the navy's

Anton Ludwig Fahnehjelm introduced modern diving dress to Sweden in the 1840s. He knew of *Vasa*'s position and sought permission to dive on the wreck. He later claimed to have dived there, but there is no independent evidence of his visit.

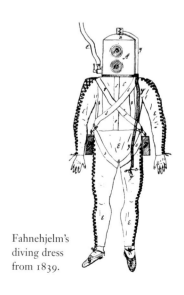

Fahnehjelm's diving dress from 1839.

LOST BUT NOT FORGOTTEN 167

Notation in the register of the Stockholm municipal authority for August 1920 of an application by the Olschanski brothers for permission to recover shipwrecks in Stockholm harbour. They were probably most interested in the cannon and waterlogged oak to be found on *Vasa*.

Commander Lenny Stackell (left) and Simon Olschanski were two of the leading people in the group interested in salvaging *Vasa* in the 1920s. The photo, probably from 1921, shows them with objects recovered by the Olschanskis from the wreck of *Riksäpplet* (sunk 1676), which they dynamited that year.

diving school was established on nearby Kastellholmen, the wreck of *Vasa* made a convenient training site.

In 1920, a salvage firm run by the Olschanski brothers of Oskarshamn applied for a permit to salvage ships lying between Beckholmen and Tegelviken, where *Vasa* lay. They had worked on other 17th-century wrecks, dynamiting them for the bronze guns and waterlogged "black oak," which was popular as a furniture and panelling material. At the same time, articles appeared in Stockholm newspapers by a noted historian, Nils Ahnlund, and a naval officer, Lenny Stackell, discussing the loss of *Vasa* in the harbour in 1628 and speculating on what might still be aboard. Permission for the salvage was refused, because the harbourmaster did not want a diving float moored in the main shipping channel, and *Vasa* escaped being blown up.

After Stackell retired, he moved to the small town of Dalarö, on the coast south of Stockholm, where one of his neighbours was a young man interested in old ships. This young man, Anders Franzén, became a fuel engineer for the navy during World War II, and in his spare time, he studied the history of the Swedish navy, especially the ships of the 16th and 17th centuries. In the early 1950s, his interest focused on twelve ships lost between 1523 and 1676. The wrecks of three of these had been found and salvaged in the 1870s and 1920s, and Franzén set himself the task of finding the other nine. Some of these ships had been lost in great battles, others sunk in storms or destroyed by fire, and one had foundered in the middle of Stockholm harbour on the first leg of its maiden voyage. He established contact with the naval history department of the National Maritime Museum and haunted the national archives, looking for clues to the locations of these ships. He persuaded the navy to let him use their boats and other equipment and to loan their divers to investigate potential sites. In many ways, he was the last link in a chain of naval tradition stretching back through Fahnehjelm and Bechstadius to the 17th century, keeping the memory of these lost ships alive.

In 1954 he began searching systematically for two ships, *Vasa* in Stockholm harbour and *Kronan* (the successor to Hein Jakobsson's *Kronan*), sunk in the open Baltic off the island of Öland in a battle with the Danes in 1676. In Stockholm, he dragged the bottom with anchors to find obstructions projecting above the mud and then sent down a small coring device he had had made by a local metalworking firm. This could retrieve a sample of the obstruction, showing whether or not it was wooden and thus potentially an old ship. For the first two years of his search he motored back and forth along the south side of the shipping channel near Tegelviken, opposite Beckholmen. This area had been suggested by the historian Nils Ahnlund as the likely site of the sinking. He found blasting rubble, scrap metal and other leavings from Sweden's recent industrial past, but no wooden ships. Things were about to change. In the winter of 1955–1956 he received important new information, and met a man destined to be a central figure in the adventure that was about to unfold.

At Dalarö, the Stackell family were neighbours of the Franzéns, and the young Anders Franzén had the chance to hear about the wrecks Lenny Stackell had been involved with. Just before his death, Stackell was present for the first dives on the wreck of *Vasa* in 1956. Franzén's hut is now a listed building.

Salvage and excavation

1956

On the 4th of September, Per Edvin Fälting sank steadily through the darkening water toward the harbour bottom 32 metres below. There was little visibility, just a swirling brownness in front of the faceplate on his helmet, but after more than 20 years of diving in Swedish waters, it did not bother him. He was used to "seeing" the underwater world with his hands, finding his feet on shifting bottoms, and working in tight, dark spaces. There was a gentle hiss from the air being pumped into the helmet from the surface, the stale smell of rubber and sweat from the waterproof canvas suit he wore, and a roaring, bubbling noise whenever he leaned his head back into the knock valve, to purge the helmet of exhaled air. This periodic tripping of the valve was instinctive in experienced divers, and other navy personnel joked that divers could always be identified by this tick, jerking their heads backwards even when not wearing diving dress.

He landed on the bottom and immediately sank slightly into the soft mud. The buoyancy of the air in his suit cancelled out much of his weight. Moving in this ooze was one of the challenges of working in the harbour, and it stirred up clouds of fine sediment, which destroyed whatever visibility there might have been. He had always had a good sense of direction, and was careful not to spin or twist around his umbilical, keeping the paired air and telephone lines clear. He knew he was to one side of the target. The sampling they had done from the surface in August showed that there was a wooden object down here several metres high and at least 20 metres long, but probably bigger. He felt for the current, turned slightly to adjust for it, and struck out across the bottom.

He had waded only a few metres through the muck when he came up against a vertical wall. It was a little soft and slimy on the surface, but firm underneath, typical of wood that had been under water for many years. The surface leaned away from him, and had raised ledges running parallel to the harbour bottom and projecting outward from the main surface by about 10 centimetres. He knew immediately that it was the side of a ship, and the ledges were wales, extra-thick

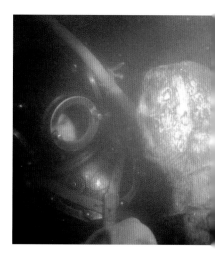

A diver face-to-face with a knighthead by the foremast. Visibility was rarely good enough for photography, and this is one of only a handful of underwater pictures taken during the salvage work.

Media attention on the lifting of *Vasa* from her resting place was intense, and publicity requirements often drove the scheduling of significant steps in the complex process of salvaging the ship.

Anders Franzén re-enacts finding *Vasa* for a press shoot. These photos led to the popular image of a man alone in a small wooden rowboat, when in fact he often borrowed motorboats from the navy in his search for *Vasa*.

planks strengthening the construction. He reported this to the tender on the surface, speaking into the microphone mounted in his helmet. He worked his way along the surface for a little way in each direction, and quickly came to an opening in the side, and then another a few metres farther along.

"There are some square openings here, maybe a metre wide – the first one is closed. The side goes higher than I can reach. I am climbing up, take in my slack." The tender on the surface pulled in some of the umbilical, so that a dangerous loop would not form, and he could slow the diver's fall if he came loose from the object he was climbing.

Fälting set one lead-booted foot on a projecting wale, grabbed the edge of the nearest opening, and began to pull himself up. After a step he could stand on the edge of the opening and reach up, to more horizontal ridges. He continued upward, until he was several metres above the bottom, and then followed the ridges to one side. He soon came to another square void, and then another, and the side continued above him.

"There is a second row of openings above the first. This has to be an old ship, very big, with two rows of gunports." On the dive boat, he knew that smiles would be shared between Anders Franzén, old Commander Lenny Stackell and the others. There was only one big ship with two gundecks sunk in this harbour, and Anders had known it.

DISCOVERY

By the autumn of 1955, Anders Franzén's two-year search for the wreck of *Vasa* had produced nothing of substance, but he was now reasonably sure that the ship did not lay on the south side of the harbour at Tegelviken. He went back to the archives, in search of more information, and found an original report that the ship had sunk near "Bleekholmsuddan," which had to be the island of Beckholmen, on the north side of the harbour opposite Tegelviken. He would have to shift his search in the coming spring.

At about the same time, he was given the results of extensive soundings taken around Beckholmen that autumn. The city was considering building a bridge across the harbour at its narrowest point and had investigated the same area that interested Franzén. Their soundings showed a distinct bump in the bottom about 120 metres off the mouth of the Gustav V drydock on Beckholmen. The bump looked to be about 50 metres long and 6 metres high, with sharply defined edges. When he inquired about it, he was disappointed to hear that it was blasting rubble from the construction of the drydock in the 1920s. Still, the possible construction of the bridge lent a new sense of urgency to his search.

During the winter, Franzén met the civilian diver in charge of the navy's dive locker, Per Edvin Fälting. He had learned to dive at the navy dive school at Kastellholmen, in the middle of Stockholm harbour, in 1932, and was probably the most experienced salvage diver in Sweden. Franzén asked him about the bump off Beckholmen, and Fälting told him that it was not from the construction of the drydock. That rubble had been dumped elsewhere. The bump must be something else, and was probably worth investigating.

Franzén's coring device, made by Palmstiernas Mekaniska Verkstad, suspended over one of the holes it probably made when Franzén and Fälting sent it down on 25 August, 1956 (the other hole has not been found).

SALVAGE AND EXCAVATION 173

Dive boss Per Edvin Fälting advises a navy diver about to enter the water. His steady hand and down-to-earth manner kept things running smoothly on the dive barge throughout the salvage.

Fälting offered to help in the search and suggested a new type of drag, a light anchor on wire rather than the heavier anchor Franzén had been using earlier. They continued to use Franzén's coring device, a heavy weight with a punch welded into one end and fins on the other. When dropped over the side, it would fall straight down and cut a plug out of whatever it hit. Using a hand reel, Franzén could pull it back up to check the plug.

During the summer of 1956, they trawled back and forth in front of Beckholmen. There was a considerable amount of debris on the bottom, most of it coming from the three busy drydocks on the island, but on August 25th, the drag fastened in something large and solid. Sounding indicated a large, sharply defined high spot in the bottom. They sent down the corer, and it came back with a plug of black oak, wood that had been submerged for a long time. Other oak plugs had come up before, and by itself one core sample was not very meaningful. They moved the boat 20 metres along the shore and dropped the corer again. It came back with more black oak. They had found a large oak structure, not just harbour trash, and it had to be investigated.

Franzén contacted the navy and asked them to arrange for a diving inspection of the object, and on September 4th, a small dive boat anchored over the site. Fälting and another diver, Sven Persson, went down for the initial inspection. Persson landed directly on the target, in a mass of tangled, broken timbers, while Fälting landed next to it and then managed to climb up the side. Several days of diving produced more information, including the size of the find, over 40 metres long. Their reports confirmed that the object was a large, wooden ship with two rows of gunports and a mast still standing. *Vasa* had been relocated.

The find was reported to the navy and the National Maritime Museum, and a small organization, the *Vasa* Committee, was formed to investigate the find further. Part of the committee's brief was to determine if the wreck could be raised, to become the centrepiece of a museum. This was Franzén's idea, and his contribution to the project was not only relocating the wreck, but his ability to recruit others to his vision of what the project could become and to convince major institutions like the navy to commit the resources necessary for the recovery of the ship. No one had ever tried to raise a ship so old before, and no one knew if it would be possible, but Franzén's enthusiasm was infectious.

An initial diving campaign was carried out in the autumn, under the overall

direction of Commodore Edward Clason, the commander of the Stockholm navy yard. He controlled all of the navy's diving and salvage assets, and had an interest in history. Fälting was put in charge of the diving, which initially created some problems in chain of command. The divers were navy ratings, while he was a civilian, but this was eventually solved by changing his job title to "dive boss."

One of the first tasks was to remove the stump of the foremast, which still stood 11 metres above the hull. It was a hazard for the divers' umbilicals and any other gear put into the water. The first divers to look at it reported the figure of a king, with the initials "IR," painted on the mast near deck level, and several others saw it when the first underwater television camera in Sweden was taken down to the site. It was thought to represent King Johan III (reigned 1568–1592), and caused speculation that this might be an earlier ship than *Vasa*, or that the mast was reused from an earlier ship. In any case, the fragile painting disappeared completely as the mast was raised.

Fälting and the "Tunnel Gang" take a coffee break during the salvage work. Fälting became a popular figure and later appeared in advertisements for coffee and other products.

Divers took measurements of the hull and explored the structure, trying to determine its overall condition and if it would hold together. They even began to tunnel under the ship, to examine the bottom and to test if it would be possible to pass lifting cables under the hull. They raised a few more objects, including a finely carved sculpture of a lion's snarling face with traces of red paint still visible in the mouth. This find became an unofficial symbol of the project and raised expectations of what else might be found in and around the ship.

Meeting in the winter, the *Vasa* Committee decided that the ship could be raised using conventional salvage techniques, and recommended that the navy, the Maritime Museum and the National Heritage Board establish a formal project to do so. A new organization was formed, the *Wasa* Board (using the 17th-century spelling of the ship's name), with Prince Bertil, the king's younger son, as chairman and Clason in command of the technical work. By this point, Franzén had persuaded one of the largest salvage firms in Scandinavia, the Neptune Diving and Salvage Company (usually known as Neptunbolaget), to join the project after getting the approval of its parent company, Broströms. Clason committed the navy's divers to the project by requiring that every diver would fulfil his annual proficiency examination by working for two weeks on the raising of the ship. The Maritime Museum and the Heritage Board would provide archaeological oversight and take care of any finds raised, and engaged a young chemical engineer, Bo Lundvall, as a conservation assistant.

There were repeated discussions of the salvage strategy, and many suggestions of how to raise the ship came forward. Some of these were rather far-fetched, including freezing the ship in a block of ice or filling it with ping-pong balls, but Axel Hedberg, the salvage chief from Neptunbolaget, effectively ended the discussion. He said that his company would raise the ship at no cost, but only if they used a method with which they were familiar. The find was too valuable to risk on experimental technology.

The method Neptunbolaget preferred was, in essence, the same method developed in the 15th century and attempted in 1629. The ship would be connected by cables to lifting pontoons on the surface and the pontoons pumped full of water. After tightening the cables, the pontoons would be pumped out, lifting the ship. This method had failed in the 17th century, because the lifting hardware was not strong enough and there was not enough lifting force at the

surface to break the suction of the thick clay in which the ship was embedded.

In the 1950s, steel cables would replace hemp rope, and instead of fastening anchors into the hull, divers would tunnel under the ship, so that the cables could be passed all the way around the ship. The tunnelling would have the added benefit of breaking the suction of the clay, reducing the force needed to pull the ship free of the bottom. Six tunnels were needed, and it would be necessary to reinforce the ship before it could be brought all the way to the surface.

It took more than two years to dig the tunnels, from the spring of 1957 until the summer of 1959. The upper layer of bottom sediment was soft, organic-rich black mud up to two metres thick and full of artefacts and debris. Under this was grey, glacial clay, hard and sticky. The bottom sediment was cut with a high-pressure water jet designed by Arne Zetterström. This had a main outlet facing forward and four smaller outlets facing backward. The smaller jets cancelled out the recoil of the larger jet, making the tool easier to handle, and they pushed the excavated mud and clay back, away from the cutting face. Working for an hour at a time, each diver used this to cut into the bottom. A large induction dredge,

Diorama in the *Vasa* Museum of a diver starting to dig one of the tunnels under the ship. He holds a water jet with Zetterström nozzle, and the suction dredge for removing the spoil is the corrugated hose entering the shaft to his left.

Only three cannon remained on the ship when it was relocated. This one was part of a grand public spectacle, with many spectators accommodated on modern Swedish warships, arranged in the summer of 1958 in conjunction with an international conference on maritime history held in Stockholm.

effectively an underwater vacuum cleaner, sucked up the sediment, which was carried to the surface and screened for small objects.

Each tunnel was dug in two halves. A diver would use the dredge to dig a wide hole next to the ship until he reached the clay layer. Then he would cut a shaft downward along the planking until he reached the turn of the bilge, where the ship's side met the bottom. Here, he would open up a chamber large enough to turn around in. From there, he would dig horizontally, following the bottom of the ship, until he reached the keel. In a hollow dug beneath the keel, he would push three steel rods into the clay, one straight ahead and one at 45 degrees to each side.

With the first half complete, diving shifted to the other side of the ship, where another shaft and room were dug and another horizontal tunnel driven toward the keel until one of the steel rods was found. The diver could then follow the rod to the end of the other half of the tunnel. This method was very effective and was needed after it proved difficult to align the two tunnel entrances on opposite sides of the ship.

The mud in the upper parts of the tunnels was full of objects which had fallen from the upper parts of the ship. In the stern, sculptures which had once decorated the transom were found in abundance, while the tunnels along the sides revealed gunport lids and rigging hardware. Initial digging on the port side also turned up the mainmast, which had been laid beside the ship after it was broken off, and a large wooden boat, later identified as the ship's longboat and raised in 1967. While clearing the upper gundeck, which was the uppermost preserved deck, divers encountered a large bronze cannon. It was the first of three 24-pounders left by the 17th-century salvors to be found, and was raised in the summer of 1958. At the very end of the tunnelling work, while the divers were making final preparations for drawing the lifting cables through the tunnels, they discovered an immense sculpture at the bow. This proved to be a gilded lion over three metres long and weighing over a ton, the ship's figurehead.

At last, in August 1959, all was ready. Neptune arrived with two lifting pontoons, *Oden* and *Frigg*, built in 1898. Although antiques themselves, they were refurbished and towed into place with two of the company's tugs. The lifting fleet was joined by the navy's diving support vessel *Belos* and various small craft. The lifting slings, paired steel cables 45 millimetres in diameter, were drawn through the tunnels and wooden packing inserted to protect the hull from abrasion. *Oden* and *Frigg* were pumped down and the cables connected. Between them, they had 1200 tons of lifting force. Calculations made by conservator Sam Svensson of the Maritime Museum suggested 600–750 tons would be enough, but no one knew how much grip the clay had on the old timbers.

On 20 August, Neptune began pumping the water out of the pontoons. The cables came taut, and everyone waited as the pumps continued to lighten the pontoons. At 4 p.m., with 600 tons of force pulling on the cables, *Oden* began to rise in the water. The ship was moving! The tunnels had broken the suction, and the hull had slipped free of the clay's sticky embrace.

Once the pontoons were empty, they had lifted the ship nearly five metres. The flotilla of vessels moored to the pontoons towed the ship away from its resting place, into shallower water until *Vasa* grounded again. The pontoons were pumped down again, and the ship settled back into the mud, but not as

Anders Franzén (right) with the key representatives of the two major public institutions collaborating in the raising of the ship: Commodore Edward Clason of the navy (left) and Count Edward Hamilton, the curator of naval history at the National Maritime Museum (centre).

1959

THE SALVAGE OPERATION

The method used to raise *Vasa* in 1956–1961 was essentially the same as developed in the 15th century and attempted in 1628–1629. It was well understood and offered a high chance of success. Other methods were briefly considered but discarded at the insistence of the Neptune Company's forceful operational director, Axel Hedberg.

The navy's diving vessel *Belos* was used for both lifts, under the command of Captain Bo Cassel, a keen diving historian who tested a replica of the diving bell used to salvage the guns in the 1660s.

For the final lift in April 1961, the pontoons were fitted with hydraulic jacks to pull the ship directly to the surface. They were kept apart by spreaders to make a space for the hull.

The lifting cables were doubled lengths of multi-strand steel cable, specially made for the salvage.

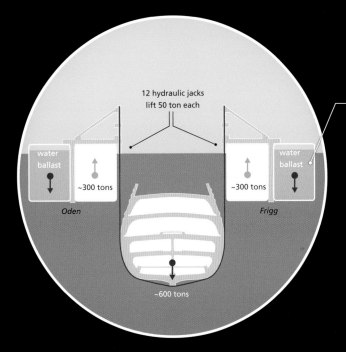

The outboard tanks in the pontoons had to be kept full of water to prevent them capsizing under the load, so only half the lifting force was available. This meant that the upper gundeck had to be cleared to reduce the weight of the hull.

Tunnels for the lifting cables began as large pits in the loose mud of the bottom but became narrow shafts in the hard clay underneath.

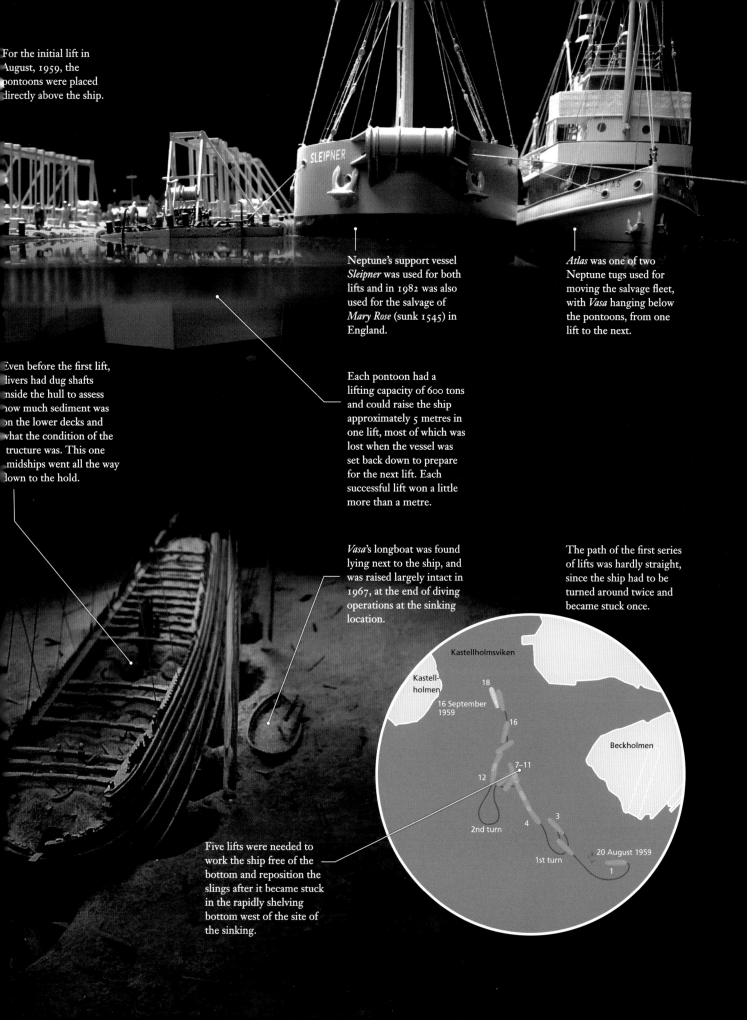

For the initial lift in August, 1959, the pontoons were placed directly above the ship.

Neptune's support vessel *Sleipner* was used for both lifts and in 1982 was also used for the salvage of *Mary Rose* (sunk 1545) in England.

Atlas was one of two Neptune tugs used for moving the salvage fleet, with *Vasa* hanging below the pontoons, from one lift to the next.

Even before the first lift, divers had dug shafts inside the hull to assess how much sediment was on the lower decks and what the condition of the structure was. This one midships went all the way down to the hold.

Each pontoon had a lifting capacity of 600 tons and could raise the ship approximately 5 metres in one lift, most of which was lost when the vessel was set back down to prepare for the next lift. Each successful lift won a little more than a metre.

Vasa's longboat was found lying next to the ship, and was raised largely intact in 1967, at the end of diving operations at the sinking location.

The path of the first series of lifts was hardly straight, since the ship had to be turned around twice and became stuck once.

Five lifts were needed to work the ship free of the bottom and reposition the slings after it became stuck in the rapidly shelving bottom west of the site of the sinking.

Interest in *Vasa*'s actual salvage on the 24th of April 1961 was enormous. Edward Clason (in trench coat farthest from camera), the driving force behind the salvage operation, watches the ship come to the surface during the final stages.

deeply as before. The first metre of the wreck's journey to the surface had been accomplished. The cables were tightened again, and the pontoons pumped out, lifting the ship so that it could be towed farther toward the shallower water to the west, in Kastellholmsviken. Each lift won another metre or so, and in a series of 18 lifts over four weeks, the ship was slowly moved into the shelter of Kastellholmen and positioned at a depth of 16–17 metres.

There were some complications along the way. While Svensson had calculated the total weight with some accuracy, the ship was much heavier at the bow than anticipated – towing it bow-first caused it to plough into the bottom. Extra lifts were required to reposition the slings, as they tended to slide along the hull, and for a two-week period, the ship was effectively stuck and a number of lifts were needed to work it free of the bottom.

Vasa spent the next 19 months on the bottom of Kastellholmsviken, while divers prepared the ship for the final lift. Winches on the pontoons would haul the ship the rest of the way to the surface, where it would be pumped out. This would only work if the old hull could be made watertight. The stern was heavily damaged by the anchor dragged through it in the 19th century, the bow was open, the gunports had lost their lids when the iron hinges rusted away, and there were thousands of empty bolt holes throughout the hull.

The navy divers who had done the bulk of the tunnelling were awarded medals and returned to their normal routine. The remaining work was accomplished by a small team of commercial divers under Fälting's leadership. They built a new, temporary transom of wood mounted on steel stanchions bolted to the hull and a similar structure at the bow. Special gunport covers were designed and clamped in place, and thousands of tapered wooden plugs were driven into bolt holes.

There were concerns about the total weight of the hull and about its stability once it reached the surface – it would not do to get the ship up, only to have it heel over and sink again. It was necessary to lighten *Vasa* as much as possible, since the lifting force available in the final stage would only be about 600 tons. Divers spent 1960 clearing as much as they could of the upper gundeck, which was over a metre deep in mud and debris in places. They threw overboard tons of blasting rubble, dumped on the ship after it sank, and recovered over a thousand artefacts, the first real indication of how rich the interior of the ship would be. These included the first human remains, coins, personal possessions, tools, and many items of ship's equipment. The divers recorded where each object was found and Maritime Museum staff catalogued the finds. The divers also tried to move some of the mud farther down into the ship, using water jets to push it into the hatches in two places. This disturbed some of the finds, including three human skeletons, but the total area affected was not very large.

The ship's structure, while it seemed very solid, might not be able to tolerate the greater forces of the final lift. The actual degree of decay of the timber was unknown and the effect of the loss of all of the iron nails and bolts could only be guessed. When the ship reached the surface, it would lose the buoyant support of the water as the hull gradually emerged. In anticipation of this, the last lift in 1959 had been used to replace the original lifting cables with heavier double cables. The divers inserted new mild steel bolts in some of the original bolt holes and installed several steel tie-rods across the ship to make sure that the upper hull would hold together.

The final work in the spring of 1961 was hectic, as a lifting date had been announced and there was great anticipation in Sweden and the rest of Europe. Neptune returned with the pontoons *Oden* and *Frigg*, which were modified by the installation of massive hydraulic jacks. Neptune also brought a commercial salvage support vessel, *Sleipner*, for the work of hooking up the cables, installing spreader bars between the cables to keep them from crushing the hull, and making the final preparations for the lift.

On April 23rd, 1961, all was ready. Divers reported that the cables were clear of obstructions, the workers on the pontoons had tested the jacks, and it was time to begin. The cables were slowly drawn in, and *Vasa* lifted free of the bottom. Over the course of the day, the hull was lifted toward the surface until it could just be seen. All was in readiness for the final lift on the morrow. In fact, the ship may have been a little eager – one of the eroded timbers peeked above the surface, and was quickly covered with an upturned coffee tin (it would not do for the ship to arrive early) until the cables could be eased slightly, lowering the ship back down a little.

People began to line the shore within sight of the lifting fleet early on April 24th. The day was sunny but cool. The harbour was full of ferries and small boats jammed with spectators, as well as police boats trying to keep some sort of order. A vessel had been chartered for the world's press, and cameras were in place to

Six hydraulic jacks were mounted on each pontoon and the ends of the lifting slings were connected to chains fixed to the jacks for the final lift. The ship could thus be lifted directly, without repeated filling and emptying of the pontoons, which would not have worked once the ship was free of the bottom.

SALVAGE AND EXCAVATION 183

Per Edvin Fälting and Anders Franzén step aboard *Vasa* at last.

broadcast the lift live to the rest of Europe. Edward Clason, Axel Hedberg and Per Edvin Fälting arrived early, as did Anders Franzén. Clason and Hedberg were in suits and topcoats, while Fälting had on a clean, white sweater and the battered captain's hat he had been given by a well-wisher early in the project. He had promised to wear it until the ship was successfully raised, and was as good as his word. Franzén wore the threadbare windbreaker he always wore to the dive barge, much to the consternation of the publicity staff.

Despite all the months of preparation, no one was entirely certain of what would happen, if the old timbers still had enough strength to support their own weight out of the water together with over a thousand tons of mud and artefacts.

Tension was taken on the jacks at 8:30 a.m. and within 30 minutes the first timbers could be seen. Before long, five weathered faces, the carved human heads on the rigging bitts standing around the masts, gazed back at the thousands watching from the shore. Once a solid piece of the starboard side was above the water, Franzén and Fälting rowed over to the ship and climbed aboard, the first people to walk *Vasa*'s decks without diving gear in 333 years.

By midday, the upper edge of the ship's side was clear of the water all the way around. Lifting continued through the rest of the day, until the upper deck beams were clear of the water. After some preparations on April 25th, three enormous submersible pumps were lowered down the hatches on the 26th. Working at thousands of litres per minute, these pumps began to empty the ship of water. The tension began to go out of the lifting cables, and the water level inside the ship sank. The interior of the upper gundeck came into view, a jumbled mess of timbers, mud, and debris, with openings along the centreline leading down into the decks below.

Pumping continued for the next ten days. Gains were impressive at first, but slowed as more of the ship emerged from the water. Divers were sent into the water to inspect the hull and to plug leaks. The traditional trick of dumping sawdust into the water and watching where it was sucked in was tried, and with the hull free of the bottom it was possible to plug more bolt holes.

In early May, the ship and lifting pontoons were towed back to Beckholmen, to the entrance of the Gustav V drydock. Inside, a purpose-built reinforced concrete pontoon was waiting. This would provide a mobile, permanent working platform for the ship while preservation work was carried out. All that was required was to get the ship into the drydock and onto the pontoon. The entrance to the dock was too narrow to allow *Oden* and *Frigg* to accompany the ship, so *Vasa* would have to float in by herself.

On May 4th, with pumps going flat out to keep up with the water leaking into the hull and Per Edvin Fälting standing on the stern, *Vasa* cast off from the lifting pontoons which had brought her to the surface. She sat low at the bow, listing heavily to port and her lower gunports were still underwater, but she was afloat on her own keel for the first time in over three centuries. Slowly, ponderously, delicately she was warped into the drydock.

Once in the dock, the great gates were closed and preparations made to move the ship onto the pontoon, which was grounded in the bottom of the far end of the dock. She was still too low in the water to get up onto the keel blocks, so four

Divers were sent into the water once the ship reached the surface to check the lifting slings and to look for leaks or damage to the hull.

A conservator carries out preliminary cleaning of the figurehead lion, raised in 1959. Storage for raised material was initially rudimentary, but eventually occupied a barge and several caves on nearby islands.

inflatable lifting bags were placed under her sides to offer a little extra flotation, enough to clear the pontoon. She was carefully manoeuvred over the keel blocks and lowered gently onto them. Fälting made one last dive to make sure that all was in order, and reported that the keel was skewed very slightly on the blocks, but safely supported over the whole length. Props were set up between the ship and the sides of the dock, to level the hull and prevent it from toppling over, and the dock was drained. As the water receded, more and more of the ancient hull came into view, until *Vasa* stood revealed from keel to railing, a sight no one had seen since the ship was launched in the spring of 1627.

DIGGING IN THE RAIN

It was expected that the ship would be full of objects, the tools, guns, ammunition, and equipment of a 17th-century warship and its crew, and much thought had been given to how to handle this material. There were discussions during the salvage work about whether it would be better to have the divers excavate the interior under water, thus lightening the ship as much as possible before raising it, or to wait until the ship was recovered, when it would be much easier to get into the hull. The engineers believed that the ship would hold together if it was raised still full of mud and artefacts, as long as the weight in the upper part could be reduced, but it would be necessary to empty the ship as quickly as possible once it was on the pontoon. The decision was taken to excavate at the surface, after the upper gundeck had been partially cleared prior to lifting.

Per Lundström, an experienced archaeologist working for the National Heritage Board, was engaged by the *Wasa* Board to direct the excavation of the interior. He recruited nine young archaeologists, as well as a draughtsman and a photographer. The archaeologists came aboard the day the ship was raised, clearing the few artefacts still lying on the small patch of the upper deck surviving on the port side towards the bow. They also had their first glimpse of what lay in store for them. The upper gundeck was the highest surviving deck, and probing by the divers had revealed three decks below it, each buried in a thick layer of mud, finds and debris. The divers had cleared the central part of the upper gundeck, but not all the way to the deck planking, leaving a compacted layer of artefacts. Deep piles of collapsed timbers still covered the ends of the deck, and the shanks and flukes of iron anchors could be seen protruding from the muck.

They started excavation in earnest two days later, in the stern, and worked there until the ship was taken into the drydock. Here lay the remains of not only the great cabin, but the two decks above it, mixed with blasting rubble. Parts of the bulkhead separating the cabin from the steerage were still standing, and fragments of the panelling which had once lined the sides of the cabin could be seen in the mud. It was a jumbled mess, probably the most complex part of the entire ship to excavate, and a challenging start. It was here that the excavation strategy was tested.

On the orlop, archaeologist Sven Bengtsson dismantles a chest in the carpenter's storeroom after removing its contents.

SALVAGE AND EXCAVATION 187

Examination of a sculpture of an armoured knight from the stern. The project bought up most of the second-hand bathtubs in Stockholm for temporary artifact storage.

Lundström was faced with a unique archaeological challenge: an intact four-story structure with nearly all of its original contents in place. The working conditions did not make it any easier. The ship had to be sprayed with cold water more or less constantly to keep it from drying and distorting, so the archaeologists worked in a steady rain, water dripping down the collars of their foul weather gear despite creative attempts to keep it out. Everything was locked in a layer of sticky, black, stinking mud. The mud also hid the structure of the deck, concealing possible openings or weak spots, and the engineers could not be sure how stable the ship was – it might collapse under its own weight at any time. There was constant pressure on the archaeologists to work as quickly as possible so that the hull could be lightened and stabilized. As they worked their way down into the hull, there was less and less headroom, so crouching and crawling were added to the discomforts. Within minutes of starting the working day, all were covered in mud.

On board, the excavation proceeded from the upper gundeck down, each deck cleared in turn as a separate excavation. The mud was gently washed away with fine sprays of water. The layers of collapsed timbers and larger objects, such as gun carriages, had to be taken apart carefully, like a giant game of pick-up sticks. Each artefact had to be lifted free of the mud, its position in the ship noted. As one team of archaeologists recovered the objects, another began cataloguing them. Each artefact was assigned a unique find number and marked with a stainless steel tag. Its location, dimensions, and distinctive details were written down in the find register, bound books of printed forms on water-proof paper, and then the object could be carried off the ship and placed in a tank of water to await conservation.

The wooden objects could not be allowed to dry out, or they would shrink and crack. Metal objects had to be kept wet as well, or they would start to corrode. A new conservation facility was under construction on the island of Beckholmen, but until this was ready artefacts were stored in hastily constructed tanks and old bathtubs out in the open and in caves which had been blasted out of the bedrock of Beckholmen by the military earlier in the century.

From the stern, the teams worked forward to the steerage, where the multiple layers of collapsed structure ended. They left the lowest layer and moved to the bow, working aft along the gundeck. The empty carriages left behind when the cannon were salvaged in the 1660s lay tossed haphazardly about, a giant's

toys abandoned in mid-game. Among them were chests and barrels full of the crew's shoes, clothing and other possessions, as well as the remaining parts of the human skeletons found by the divers clearing the deck in 1960. As they cleared the last layer in the steerage, they found the first complete skeleton (later called "Filip"), lying near the whipstaff.

The lower gundeck was less disturbed, protected by the deck above. The gun carriages still stood at their ports, more barrels and chests were intact, and there was less loose wreckage from the rigging and upper deck. The objects found here were mostly material associated with the guns and the personal possessions of the crew. It was possible to clear the deck relatively quickly and to probe farther down into the hull.

While they dug, the technical staff trying to stabilize the ship were increasingly concerned about the weight they were asking the waterlogged timbers to support. They were especially worried about the many tons of ballast and the deep

The lower gundeck, looking forward on the port side. As the gundeck was cleared, the gun carriages were left in place, still standing at their gunports. Once the space had been photographed, the carriages were dismantled and removed for conservation.

The ballast, over 10,000 individual stones, was initially stored at the conservation facility on Beckholmen, except for a small amount used in the *Vasa* grave at the naval cemetery, but was moved in 2009 to the bottom of the drydock basin in front of the *Vasa* Museum.

layer of mud in the hold, which might cause the hull to crack. They persuaded the archaeologists to bypass the orlop, the deck directly beneath the lower gundeck, and to clear the hold next.

The bottom of the ship was still watertight, which had allowed the ship to be refloated, but was now holding in all of the water being sprayed inside the ship. Attempts to pump the water out were thwarted by material clogging the pumps, and eventually three holes were cut in the bottom, near the keel on the starboard side amidships and near the ends. These drained through screens, to catch any objects.

The sight that greeted the archaeologists once the water was gone was disorientating. The ballast had shifted when the ship sank, so that a mound of stones and artefacts sloped up towards the port side. The ballast deck, loose planks laid over the stones, was scattered through the hold, and the space was divided into a warren of small compartments by the partially collapsed remains of bulkheads. At the centre stood the galley, with most of its brick floor and walls still in place at the bottom of the main hatch, just before the mainmast.

The ballast stones had to be removed as quickly as possible, and several solutions were proposed. Lundström suggested cutting a hole in the side of the hull and passing the stones out, but the engineers felt this would damage the structure too much. Eventually, the ballast was taken out by reversing the process used in loading it. The stones were carried to the galley and loaded onto pallets. These were taken up by a lift built in the main hatch and then moved ashore. Each load was weighed, for a final total of 119.7 tons of stone.

Much of the hold proved to be empty, as the ship had not been fully provisioned when it sailed. One small compartment was full of barrels of meat and the compartments at the stern produced some of the highest-status objects, such as pewter plates and a complete backgammon set. Towards the bow, the anchor cables were found in large coils. Some of the cable could be removed, in short sections laid into plastic roof guttering, but much of it was too decomposed and had to be taken out in buckets. Some of this decayed fibre was later used to make rag paper, sold as a souvenir. One coil was removed as a solid lump, and was found to contain most of the skeleton of a man (later designated "Johan").

The orlop was the last space to be investigated. Because of the low headroom, less than a metre under the beams in places, the archaeologists worked on their

knees. More spares for the ship, such as a tiller over 9 metres long, and some personal possessions were found here, but the most spectacular find did not at first look like much. A small compartment at the stern was nearly filled by a muddy mound of rope and decayed fibre. Investigation showed that this was a pile of sails, folded and tied up with cord as they had been delivered from the sailmaker. Getting it out of the ship proved one of the most difficult challenges of the excavation. It was eventually cut into three separate pieces, with the matching ends of ropes at each cut labelled. Thin sheets of steel were carefully slid under each part and then dragged forward onto lifting trays, which could then be taken out through the main hatch.

With the sails removed, the excavation was complete. Over 30,000 objects had been recovered in the course of five months. While the interior was being cleared, the shoring up of the hull had been completed as well, and the ship on its pontoon moved out of the drydock and around the corner of Beckholmen to a mooring. Here, a prefabricated aluminium house was erected over the ship to

The majority of the objects associated with *Vasa* are now stored in a climate-controlled magazine in the pontoon under the ship, although unwieldy objects, such as the mainmast core and most of the gun carriage parts, are held at a remote site shared by the museums of Stockholm.

By far the most tedious and challenging task undertaken by the conservators was the unfolding and stabilization of *Vasa*'s sails, which took nearly a decade.

protect it during the conservation treatment to come. In November, the pontoon was towed to the new *Wasa* Shipyard, about 200 metres to the northeast on the island of Djurgården. The ship would be restored and preserved here, in a temporary museum, while a permanent museum was built nearby.

At every stage, the *Vasa* Project presented new and unique challenges. The problem of conserving a wooden object over 45 metres long and weighing hundreds of tons was the most pressing once the ship was raised. It had been discussed at length during the salvage project and trials had been carried out with traditional treatments as well as innovative methods. The water filling the wood cells had to be replaced with something that would maintain the shape of the cells, so that they did not collapse as the water evaporated. Paradoxically, the problem was complicated by the good condition of the timbers. While the outer 10–20 millimetres were decayed and fragile, the wood beneath was still quite sound. The decayed outer layer was absorbent and would take up a preservative, but the sound core was nearly impervious to most treatments. The different rates of shrinkage which would result from most traditional treatments would cause the outer surface to crack and flake off. Established treatments that worked on small objects were not practical on something so large, or were only suitable for objects which were decayed all the way through. A new chemical developed in Sweden for the forest industry, polyethylene glycol (PEG), showed great promise and was chosen as the primary treatment for the hull and the majority of the wooden small finds.

Objects were best treated by immersion in a tank, but this was not possible with the hull. Instead, a solution of PEG in water was sprayed, at first by hand, but this proved impractical. An automatic system was installed in 1962, and this ran for 17 years. One of the challenges with a new treatment is knowing when it is finished, and there was considerable debate about when to stop spraying and start the long process of drying. After spraying stopped in 1979, the ambient humidity in the ship hall was slowly lowered towards the target of 60 percent over the following decade.

The ship was not simply a chemical challenge, it was also a giant jigsaw puzzle. Five seasons of diving at the site of the sinking followed the raising of the hull, and the small team recovered thousands of loose timbers and sculptures from the beakhead and sterncastle, including the complex and heavily decorated quarter galleries. These had to be remounted on the ship, as did the collapsed decks found in the upper gundeck. The intact hull also had to be stabilized. The original iron bolts which held the deck structure together and reinforced the hull had corroded away completely. Over five thousand new, mild steel bolts were driven into the original holes, after hydraulic jacks pushed the sagging hull back into something like its original shape. The hull needed a more permanent support structure to replace the temporary shoring set up when the ship was first placed on the pontoon, so a series of steel cradles was erected under the hull.

The reassembly of the collapsed beakhead and stern created its own challenges. The breakage of key strength elements in the stern meant that the structure could not support itself. A stainless steel subframe was built into the upper transom, with arms running forward to tie the fragile timbers into the heavy, intact structure of the ship's sides. The restoration brought a few surprises as well. The pontoon and the house over the hull were designed and built before the loose timbers of the beakhead and sterncastle had been recovered, so the overall dimensions were based on estimates of the ship's original size. Reassembly showed that the beakhead was longer and the sterncastle higher than anticipated, and would not fit into the existing building. Reconstruction work stopped for a year and half while the house was expanded.

The final phase of the restoration and conservation was the rigging. The original foremast and mainmast, as well as many of the smaller parts of the rig, were found in and around the ship. After the ship was moved into its permanent home in 1988, it was possible to step the masts and rig the ship, as she would have appeared in winter ordinary in the 17th century.

One planned phase of the restoration could not be completed. From the beginning of the project, the ultimate goal had been to restore the ship completely, to place the conserved finds back in their original places, as if the ship were just leaving the quay in August of 1628. The public could then walk through the ship, to experience a complete 17th-century environment. By the early 1970s, it was clear that the ship could not be made safe for large numbers of visitors (it would have been necessary to cut fire exits in the hull, for example), and the ancient timbers, as sound as they were, would never stand up to the traffic of hundreds of thousands of people every year. The ship had to remain closed to the public.

In the 1990s, with the ship in the new museum, it was possible to step the lower masts and rig the shrouds and stays, largely with original deadeyes, under the direction of Olof Pipping. Rigger's apprentice Mikael Hinnersson sits on the tenon at the mainmast head, more than 20 metres above the deck.

SALVAGE AND EXCAVATION 193

Vasa today

2011

As soon as the ship reached the surface, it was an international sensation. There was enormous pressure to open the ship to visitors, and a steady stream of VIPs came through during the excavation. Planning for a temporary home for *Vasa* had begun well before the ship was raised and public access to the find as soon after the salvage as possible was an important goal. While the archaeologists were clearing the interior of the ship, construction crews were busy building the *Wasa* Shipyard, a semi-industrial facility on the island of Djurgården for the conservation and restoration of the hull and a state-of-the-art conservation laboratory on Beckholmen for treating the thousands of small artefacts recovered from the ship. The ship house, built over the hull on its pontoon, included galleries at several levels so that visitors could watch the ship being restored and conserved. The shipyard included offices and an exhibition area where newly conserved objects could be shown.

It was expected that it would take about a decade to conserve and restore the ship, but the ship eventually spent 27 years at the shipyard. From the day the shipyard opened to the public, November 30, 1961, it was one of the most popular attractions in Stockholm. In the first full year, hundreds of thousands of visitors were happy to peer through the curtain of cold water and polyethylene glycol (PEG) being sprayed on the old timbers. Visitor numbers dipped down towards a quarter of a million in the mid 1960s, but rose again steadily in the 1970s, as the restoration began to bring the ship back to something closer to its original appearance. By the 1980s, visitors had stabilised at about a half a million each year, most of them from outside Sweden.

Planning began very early for a permanent home. In 1971 agreement was reached that the new museum would be built somewhere within the old Galley Wharf (Galärvarvet) area on Djurgården, where the 18th-century galley fleet built to protect the approaches to Stockholm was housed in the winter. This stretched from the bridge onto the island at the western end down to the *Wasa*

The architects of the *Vasa* Museum, Marianne Dahlbäck and Göran Månsson. They were able to follow the construction of the new museum from their office on Kastellholmen.

The *Vasa* Museum today, with its distinctive copper roof and the three masts which show the rig's original height. The façade has a striking appearance from the water, which was an important part of the design concept.

The *Wasa* Shipyard was a provisional museum but still one of Stockholm's great attractions, even though the building was so small that visitors could only get a close-up view of dripping timbers.

Shipyard site. A decade later, an international competition was announced for the design of the new museum. From 384 entries, some of them fantastic but impractical, the jury awarded two first prizes, one to a Danish entry and one to a Swedish entry. Both were invited to present detailed proposals. The Swedish design, by Ove Hidemark and Göran Månsson, was chosen as the more practical to build. Månsson and his wife, Marianne Dahlbäck, proceeded to develop the design into finished construction drawings. Their work was based on a unique, multi-angled space with an open interior, which would allow the ship to be seen from all of the exhibit areas. The original rig would be reflected in steel masts and sail outlines on the roof. The interior would feature an industrial style, recalling the ambience of the sea and a working shipyard. Climate and other systems were designed for an anticipated 600,000 visitors a year.

Månsson and Dahlbäck's design used as a central element the navy's old drydock at the Galley Wharf. It had been built in the 1870s for the navy's first generation of armoured warships but was too small for the ships which followed. It lay only a few hundred metres from the *Wasa* Shipyard and was still functional, so that the ship on its pontoon could be moved to the site with minimal complications.

Construction began in 1987, with blasting to widen the drydock slightly, and the shell was completed by the autumn of 1988. The southwestern wall over the drydock had been left open, to allow the ship to enter. While the building was under construction, the restored ship was encased in a temporary protective shell and the house on the pontoon dismantled. On a dreary, snowy day in December 1988, the pontoon was towed, once again by the Neptune Company, from the *Wasa* Shipyard to the new *Vasa* Museum (the modern spelling was chosen as the official version). The ship was carefully manoeuvred into the new building, with barely a half metre of clearance on each side, the drydock gates were closed, and the water was pumped out of the dock. It had been decided that the ship should be displayed at the same draught as when she sailed, 14 feet at the bow and 16 feet at the stern. It took 36 hours of pumping and adjusting before the pontoon settled into its final position on a forest of concrete pylons and the Neptune project director could report that the ship "had completed its only successful voyage."

The building could be finished with the ship inside, and the construction of thematic exhibits about different aspects of the ship's history could begin. The

museum was partially open in the summer of 1989, and was officially opened in the summer of 1990. Over a million people visited the museum in its first year before settling down to around 750,000 annually. After 2003, the number of visitors began to climb steeply. By 2010, over a million people a year were regularly coming to see the ship, three-quarters of them from foreign countries.

It is a remarkable tale, creating a sensation from what was a total failure. Partly this is the legacy of the heroic story of the modern salvage, partly the result of the astonishing level of preservation. Even if visitors cannot go onboard, they are confronted by an essentially intact, 17th-century object over 60 metres long and seven stories high, decorated with over 700 original sculptural ornaments. More than 98 percent of the timber in the ship as it is now displayed is original material from 1628. This includes not only the hull but the foremast and mainmast, the main top (the round platform at the masthead), and most of the blocks and deadeyes in the rigging. The only significant new material in the ship is the planking of the upper deck, the outer end of the bowsprit, and the mizzenmast. Those not interested in ships or cannon are still fascinated by the remains of

The *Wasa* Shipyard's pontoon house was a long-lived improvisation and a distinctive feature on the Stockholm skyline. Here, the newly completed structure is towed to its mooring on Djurgården in the autumn of 1961.

Vasa, encased in a temporary protective cover, is towed into its new museum in December 1988. Once inside, the drydock could be emptied and the last wall completed, sealing the ship in.

everyday life, from plain clothing and sturdy shoes to fancy clocks, a moment frozen in time.

NEW CHALLENGES

When the spraying of PEG preservative stopped in 1979, it was believed that the conservation of the ship was essentially finished. As the ship began to dry, the conservation effort focused on trying to prevent distortion of the hull and working with the building owners to improve the climate. The primary fear for the future was that chemical changes in the PEG would affect the hull. The conservation facility was gradually dismantled and many of its staff moved on to other jobs. When problems did arise, they came from an unexpected direction.

Already in the 1990s it became apparent that the ship was not stable. Cracks appeared once the ship started drying, and gaps opened between the planks as the timbers shrank. Bulging of the hull between the supporting elements of the cradle showed that the timbers were not as stiff or strong as once thought. Extra cradle elements were added. Not all of the original bolts had been replaced, and a major effort was required to add bolts that fastened the deck structure into the rest of the hull, keeping the ship from splitting open. The stern began to droop, and cables were attached to the stainless steel subframe built into the transom so that some of the weight could be suspended from the ceiling. White and yellow powdery deposits began to appear on the surface of the wood, both in the hull and on artefacts stored in the magazines.

By the beginning of the new millennium, it was clear that the conservation of the ship was not finished, but an on-going process, and stop-gap measures would not suffice for the long term. A thorough reassessment of the condition of the ship and artefacts was needed, along with a plan for preservation that reached beyond decades and considered centuries as a time scale. Wet conditions in the summer of 2000, combined with near-record numbers of visitors, caused visible changes that brought matters to a head.

Neptune succeeded in the precision task of manoeuvring *Vasa* into the new museum, and the ship's voyage was at last completed.

The powdery deposits were analyzed and found to be highly acidic. If not neutralized, they would attack the wood at the cellular level, destroying the strength of the timbers. Regular measurement of the hull with a laser theodolite showed that seasonal movement of the ship was stressing the timbers, and if not stabilized would weaken the structure's ability to support its own weight. Examination of the steel bolts, less than 40 years old, showed that they were corroding, adding to the ship's problems. Some of these changes were directly related to the unstable climate in the ship hall. Instead of a steady temperature and humidity, the air in the hall was subject to wide fluctuations during the day, as sunlight warmed the building and visitors brought in extra humidity with wet clothes and their own exhaled breath.

The most immediate challenge was to stabilize the climate, which affected both the chemical and mechanical stability of the wood in the ship and thousands of artefacts in the magazines. The original system could not handle the maximum visitor capacity of the building, 1450 people, which was being reached much more often than anticipated. It had been seriously overloaded from the moment the museum opened, so it had to be upgraded. In 2003–2004, new components were installed, with digital sensors inside and outside the ship feeding real-time data to

Maritime archaeologist Krum Batchvarov documents the framing of the hull. Research continues 50 years after the salvage, and the ship attracts scholars from Sweden and the rest of the world, eager to learn new things about the old ship.

a central computer, which could adjust the temperature and humidity in response to changing conditions in the ship hall. After it came on line, both temperature and humidity could be maintained at a chosen level within a range of just a few per cent. Measurements of the ship showed that both the chemistry and the structure settled down almost immediately. The new climate not only established a much better environment for long-term preservation but has bought time to address the other challenges.

At the same time that the climate was being addressed, a series of international research projects were carried out with partner institutions in Sweden, Denmark and the UK to analyze the chemical changes in the wood and their consequences for the structure. These studies suggested that sulphur compounds and other pollutants absorbed by the ship while it was on the bottom of the harbour were combining with moisture in the air and iron corrosion products to produce a range of acids, and that the mechanical properties of PEG-treated, waterlogged wood attacked by acids were substantially different from the properties of fresh oak.

The changing nature of the wood meant that it could not be relied on to support its own weight in the long term, and the corrosion of the bolts meant that they were making the problem worse. Analysis of the ship structure showed that the original cradle did not correspond very closely to the actual distribution of stress in the hull. The solution is a complete replacement of all of the 1960s steel bolts with new bolts in a more stable material, stainless steel, and the construction of an entirely new cradle.

Changing the bolts required the development not only of a new bolt that could be inserted without damaging the hull and could be adjusted easily, but also a method for extracting the old bolts safely.

The cradle is an even bigger challenge, as it requires a much better knowledge of how the original structure was put together. The first step has been a multi-national project to document the hull to a consistently high level, with the help of students in maritime archaeology from universities in Sweden, Denmark, the UK and the USA. The data from this project, which was completed in 2011, is combined with the results of engineering studies of the decayed wood's mechanical properties to determine where new support elements should be located. It is already apparent, both from these studies and the annual survey of the hull that the sides of the hull are not strong enough to support the decks, which were overbuilt by the shipwrights at the navy yard in 1626–1627. Not only are these heavy decks one of the causes of the ship's loss in 1628, they are a threat to its continued survival. Some sort of internal structure will be needed to lift this weight off of the sides.

The support structure is the most visible of the measures taken to preserve the ship, but it will not be the last. Wooden buildings up to a thousand years old survive in other parts of the world, weathering rain and snow in the open countryside. If we wish *Vasa* to last as long, it will take a concentrated and unflagging effort. Because this ship was the first to be conserved by impregnation with PEG, which has become the standard method for large, waterlogged wooden objects, it is the first to encounter the long-term consequences. Each generation

Conservator Emma Hocker measures pH values on a piece of *Vasa* oak. Low pH indicates potential acidic attack. Large-scale research projects have been carried out since 2003 to try to understand the chemical processes affecting the timber.

Over a million people visit the *Vasa* Museum every year. Most of them come from abroad. One can see the museum as a large display case which the visitors share with the ship, so the climate system has to cope with heat and humidity they generate.

working with this ancient ship discovers new challenges, and must rise to meet them. There is now a stable climate in the museum, and the physical support of the hull will be assured by the current generation, but what will their children and grandchildren face, so that their grandchildren can continue to enjoy the magnificent sight which greets visitors today?

The 21st century has seen not only an expanded preservation effort, but a renewed interest in understanding the ship, the people who built and sailed her, and their historical context. The museum has an active program of research, often in cooperation with universities in Sweden and abroad, recording and studying the hull and the thousands of objects found in and around it. This book is one product of that program. Even fifty years after the salvage, new discoveries are still being made virtually every week. For example, for many years it was wondered why the ship carried no navigation equipment other than sounding leads, but since 2009, the ship's binnacle, compass and sandglasses for timing the watches have been discovered in the magazines. Some of this research has shown that certain parts of the *Vasa* story that were well known and popular are more myth than fact: it can now be shown that the ship was not enlarged while it was being built, but was planned with two gundecks from the beginning. *Vasa* will continue to teach new lessons in the years to come, as each generation of scholars asks new questions, and each visitor brings a new perspective.

FIASCO OR FABLE?

Vasa was raised in a pivotal era in the development of maritime archaeology. The short period of 1956–1962 saw the discovery and excavation of well-preserved ships in Egypt (the Pharaoh Khufu's barge from c 2600 BC), Turkey (the Late Bronze Age trading ship at Cape Gelidonya and a Byzantine wine carrier at Yassıada), Denmark (the Skuldelev Viking ships from the 11th century AD), and Germany (a Hanseatic cog sunk in 1380 in Bremen harbour). Together with *Vasa*, these finds created a popular impression of what lay under the world's seas and what modern technology could accomplish. The recovery of these ships led to the creation of new museums and new institutions, which developed these successes into long-term research programs, but few of the finds which followed had the impact of that first generation of ships. *Vasa* shows that old ships have an important story to tell, one that appeals to a broad audience if it is presented in the right way.

Today *Vasa* is a major international visitor attraction, with a multilingual staff introducing people from all over the world to the history of the 17th century. It may have been an embarrassment to the Crown when it sank, but now, it still performs its intended role as a visible symbol of Sweden, a symbol of Gustav II Adolf's ambitions for himself and his country. Kings and battles are not the only attraction: the people who sailed on the ship and died when she sank, the ordinary men and women who are otherwise utterly lost to history, can have their stories told. *Vasa* is a rare window on a vanished world, a turbulent era of violence and faith, dreams and disappointment. One would have to argue that the ship is a far greater success than anyone alive in 1628 could have imagined.

A significant part of the research effort at the *Vasa* Museum takes place in front of the public and new information can be communicated directly to them by a professional corps of guides. Here rigging expert Olof Pipping makes adjustments to a 1:10 model of the ship while visitors pass by and ask questions.

Source material

Much of what can be read in these pages is based on the surviving original documents relevant to *Vasa* or the archaeological analysis of the ship and finds. Many chapters begin with a dramatized version of events from a specific person's perspective. The thoughts put into their heads are, of course, speculation, but the events described can be confirmed by historical or artefactual evidence. We cannot know what Söfring Hansson thought about losing his ship, but we can reconstruct the process of the sinking quite accurately.

Most of the relevant documents are held in the Swedish National Archives. We are fortunate that the records relative to *Vasa* do not seem to have suffered much in the fire that damaged the royal palace in 1697, but the source material is spread over several different collections due to periodic reorganizations of the archives. Much of the most useful material is in *Riksarkivet*, in the collections associated with the management of the navy yard (*Skeppsgårdshandlingar*) and the military records (*Militärsamlingar*), as well as the financial accounts (*Rikshuvudböcker*) and the registry copies of outgoing correspondence (*Riksregistraturet*). Because *Vasa* predates the formal creation of the Admiralty Board, whose records eventually were combined with the army's to form the Military archives (*Krigsarkivet*), there is less material there, with the exception of the records regarding the management of the Crown's guns, both on land and at sea, which are found in the artillery accounts (*Arkliräkenskaper*). Many of the participants in the *Vasa* story, including Söfring Hansson and the Hybertsson family, appear in the minutes of the small claims court of Stockholm (*Stockholms stads tänkeböcker*, which have been published). A final source worth consulting is the archive of the Svea court of appeals (*Svea hovrätt*), where the suit brought by Andreas Peckell against Albrecht von Treileben was referred. These feature extensive testimony from all of the participants in the salvage of *Vasa*'s guns in 1663–1665.

The archaeological source material is in the process of being studied and published as part of a long-term research initiative called *Förstå Vasa/Understanding Vasa*. The first volume of a monograph series on the archaeological material appeared in 2006 (Carl Olof Cederlund: *Vasa I: The Archaeology of a Swedish Warship of 1628*. Statens maritima museer).

The list of literature which follows is more in the nature of an extended reading list than a scholarly bibliography. It includes general works relevant to the period, as well as studies of *Vasa* and her finds.

GENERAL SOURCES, MARITIME ARCHEOLOGY

Adams, Jon, 2003: *Ships, Innovation and Social Change: Aspects of carvel shipbuilding in Northern Europe 1450–1850*. Stockholm Studies in Archaeology 24/Stockholm Marine Archaeological Reports 3. Stockholm University.

Bäckström, P.O., 1884: *Svenska flottans historia*. Stockholm.

Cederlund, Carl Olof, 1983: *The Old Wrecks of the Baltic Sea. Archaeological recording of the wrecks of carvel-built ships*. B.A.R. International Series 186, Oxford.

Cederlund, Carl Olof, 1995: The Regal Ships and Divine Kingdom. *Current Swedish Archaeology* 2: 47–85. Also in M. Acerra (ed.), *L'invention du vaisseau de ligne 1450–1700*. Paris 1997.

Cederlund, Carl Olof, 1997: *Nationalism eller vetenskap? Svensk marinarkeologi i ideologisk belysning*. Stockholm.

Cederlund, Carl Olof, 2002: From Olaus Magnus to Carl Reinhold Berch. On the background of Swedish marine archaeology and ship archaeology in the history of ideas. *Deutsches Schiffahrtsarchiv* 25.

Cederlund, Carl Olof, 2006: From Carl Reinhold Berch to Nils Månsson Mandelgren. On the concept of maritime history, (Sw. *sjöhistoria*), and its meanings in Sweden since the latter part of the 18th century. In L. Blue, F. Hocker and A. Englert (eds.), *Connected by the Sea. Proceedings of the 10th International Symposium on Boat and Ship Archaeology*. Oxford.

Jensen, O.W. 1999: *Historiska forntider. En arkeologihistorisk studie över 1000–1600-talens idéer om forntid och antikviteter*. Institute of Archeology, University of Gothenburg.

McNeill, William H., 1982: *The Pursuit of Power: Technology, Armed Force, and Society since A.D. 1000*. Chicago.

Muckelroy, Keith, 1978: *Maritime Archaeology*. Cambridge.

Åkerlund, Harald, 1951: *Fartygsfynden i den forna hamnen i Kalmar*. Stockholm.

GENERAL OR COMPREHENSIVE SOURCES ON *VASA*

Cederlund, Carl Olof, 2006; *Vasa I: The Archaeology of a Swedish Warship of 1628*. Stockholm.

Kvarning, Lars-Åke and Ohrelius, Bengt, 1998: *The Vasa, the Royal Ship*. 4th ed. Stockholm.

Landström, Björn, 1980: *The Royal Warship Vasa*. Stockholm.

AN AGE OF FAITH AND VIOLENCE

Bechstadius, Carl Nilsson, 1734: *Then Adelige och Lärde Swenske Sjö-Man ...* Karlskrona.

Glete, Jan, 1993: *Navies and Nations: Warships, navies and State Building in Europe and America 1500–1860*. Stockholm Studies in History 48:1. Stockholm.

Glete, Jan, 2000: *Warfare at Sea, 1500–1650: Maritime Conflicts and the*

Transformation of Europe. London and New York.

Glete, Jan, 2002: *War and the State in Early Modern Europe: Spain, the Dutch Republic and Sweden as Fiscal-Military States, 1500–1660*. London and New York.

Glete, Jan, 2010: *Swedish Naval Administration 1521–1721: Resource Flows and Organizational Capabilities*. Leiden.

Hocker, Fred, 1999: Technical and organizational development in European shipyards 1400–1600. In Jan Bill and Birthe Clausen (eds.), *Maritime Topography and the Medieval Town*, pp. 21–32. Copenhagen.

Lockhart, Paul Douglas, 2004: *Sweden in the Seventeenth Century*. Basingstoke.

Marinstaben, 1937: *Sveriges sjökrig 1611–1632. Särtryck ur Generalstabens Sveriges krig 1611–1632*. Stockholm.

Probst, Niels M., 1996: *Christian 4.s flåde*. Marinehistoriske skrifter. Copenhagen.

Roberts, Michael, 1953–58: *Gustavus Adolphus. A History of Sweden 1611–1632*. London.

Roberts, Michael, 1979: *The Swedish Imperial Experience 1560–1718*. Cambridge.

Wanner, Michal, 2008: Albrecht of Wallenstein as General of the Ocean and the Baltic Seas and Northern maritime plan. *Forum Navale* 64: 8–33.

Wolke, Lars Ericson and Martin Hårdstedt, 2009: *Svenska sjöslag*. Forum Navales skriftserie no 36. Stockholm.

Wolke, Lars Ericson, Göran Larsson and Nils Erik Villrand, 2006. *Trettioåriga kriget: Europa i brand 1618–1648*. Stockholm.

Zetterstén, Axel, 1891: *Svenska flottans historia: åren 1522–1634*. Stockholm.

TIMBER, IRON AND MEN

Börjesson, H., 1942: Sjökrigsmateriel och skeppsbyggnad åren 1612–1679. In *Svenska flottans historia* I, pp. 233–294. Malmö.

Cederlund, Carl Olof 1966: *Stockholms Skeppsgård 1605–1640*. Licentiate thesis, Stockholm University.

Glete, Jan, 2002: Kontrakt, bestick och dimensioner: Vasa och de stora skeppsbyggnadskontrakten under Gustav II Adolfs tid. En sammanställning av data ur kontrakt, räkenskaper och administrativt källmaterial rörande flottan. Internal research memorandum, *Vasa* Museum.

Hallenberg, Mats, 2009: *Statsmakt till salu: Arrendesystemet och privatiseringen av skatteuppbörden i det svenska riket 1618–1635*. Stockholm.

Hocker, Fred, 2004: Bottom-based shipbuilding in northwestern Europe. In Frederick M. Hocker and Cheryl Ward (eds.), *The Philosophy of Shipbuilding: Conceptual approaches to the study of wooden ships*, pp. 65–94. College Station.

Lemée. Christian P.P., 2006: *The Renaissance Shipwrecks from Christianshavn: An archaeological and architectural study of large carvel vessels in Danish waters 1580–1640*. Ships and Boats of the North 6. Roskilde.

Löfgren, Albert (ed.), 1932: Näckströmstrakten och Packaretorgsviken. *Samfundet S:t Eriks årsbok*. Stockholm.

Madebrink, Olle, 2005: *Han som byggde de stora skeppen, en uppsats om skeppsbyggmästaren Jacobssons verk*. B.A. thesis, Stockholm University.

Rålamb, Åke Classon, 1691: *Skeps Byggerij eller Adelig Öfnings Tionde Tom*. Stockholm.

Van Yk, Cornelius, 1697: *De Nederlandsche Scheepsbouwkonst open gestellt*. 2nd ed. Amsterdam.

Witsen, Nicolaes, 1671: *Aeloude en Hedendaegsche Scheeps-bouw en Bestier*. Amsterdam.

THE MACHINE OF WAR

Brusewitz, Bengt, 1985: Beväpning och övrig tygmaterial. In Jonas Hedberg (ed.), *Kungl. Artilleriet, Yngre vasatiden*, pp. 61–98. Kristianstad.

Caruana, Adrian, 1994: *The History of English Sea Ordnance 1523–1875 vol. I: The Age of Evolution 1523–1715*. Ashley Lodge.

Clason, Edward, 1964: Om Vasas bestyckning [in two parts]. *Tidskrift i Sjöväsendet* 127.11 and 127.12: 762–779 and 849–864.

Hamilton, Edward and Sandström, Anders, 1982: *Sjöstrid på Wasas tid – taktik, artilleri och eldhandvapen*. Vasa Studies 9. Stockholm.

Holmstedt, Nils, 1985: Skjutteknik. In Jonas Hedberg (ed.), *Kungl. Artilleriet, Yngre Vasatiden*, pp. 99–120. Kristianstad.

Höglund, Patrik, 2002: "Opå Wasan ähr inskepade in Julio" – om vapen, ammunition och artilleritillbehör på örlogsskeppet *Vasa*. *Marinarkeologisk Tidskrift* 25.4: 22–27.

Jakobsson, Th., 1938: *Lantmilitär beväpning och beklädnad under äldre Vasatiden och Gustav II Adolfs tid*. Stockholm.

Schultz, Johan, 2000: *Ballistik på 1600-talet. Ett samspel mellan vetenskap och teknik*. Stockholm Papers in the History and Philosophy of Technology. Stockholm.

Seller, John, 1691: *The Sea Gunner*. London.

THE SYMBOLIC WARSHIP

Fajersson, Malin (ed.), 1999: *Vasa bekänner färg*. Stockholm.

Soop, Hans, 1986: *The Power and the Glory: The Sculptures of the Warship* Wasa. Stockholm.

Tångeberg, Peter, 1986: *Mittelalterliche Holzskulpturen und Alterschreine in Schweden*. Stockholm.

THE FULL-RIGGED SHIP

Andersen, R.C., 1927: *The Rigging of Ships in the Days of the Spritsail Topmast, 1600–1720*. Salem, MA.

Cavallie, J. 1976: Werner von Rosenfeldts sjömansmemorial. *Karolinska Förbundets årsbok 1976*: 100–156.

Dalhed, Maria, 1997: Människorna kring Vasas rigg. In Katarina Schoerner (ed.), *Det seglande skeppet: Uppsatser kring en utställning*, pp. 9–18. Stockholm.

Hand, Johan, 1879: *Johan Hands dagbok under K. Gustaf II Adolfs resa till Tyskland 1620*. Historiska handlingar del. 8, N. 3. Stockholm.

Harland, John, 1984: *Seamanship in the Age of Sail: An account of the shiphandling of the sailing man-of-war 1600–1860, based on contemporary sources*. London and Annapolis.

Harland, John, 2003: *Capstans and Windlasses: An illustrated history of their use at sea*. Piermont NY and Florens OR.

Törnquist, Leif, 2008: *Svenska flaggans historia*. Stockholm.

Von Rosenfeldt, Werner, 1693: *Navigation Eller Styrmanskonsten Til Ungdomens Nytta Wed Kongl. Ammiralitet*. Stockholm.

Åhlund, B. 2002: *Historia kring våra sjökort*. Forum Navales skriftserie nr 5. Karlskrona.

A FLOATING COMMUNITY

Ahlström, Bjarne, Almer, Yngve and Hemmingsson, Bengt, 1976: *Sveriges mynt 1521–1977: The Coinage of Sweden*. Stockholm.

Hansson, Hans, 1962: Wasa och Stockholm. *S:t Eriks årsbok*: 9–24.

Kaijser, Ingrid, 1982: *Ur sjömannens kista och tunna*. Vasa Studies 10. Stockholm.

Lindblom, Irene, 2002: An analysis of the officers' storeroom in the hold of the *Vasa* and a new interpretation of a closed find in the same. In Kersten Krüger and Carl Olof Cederlund (eds.), *Maritime Archäologie heute*, pp. 323–333. Rostock.

Lindegren, Jan, 1980: *Utskrivning och utsugning. Produktion och reproduktion i Bygdeå 1620–1640*. Uppsala.

Martinsson, A., 1963: The geological provenance of net-sinkers found in the wreck of H.M.S. *Wasa* in Stockholm. *Geologiska Föreningens Förhandlingar* 85: 287–297.

Schoerner, Katarina, 2002: Om *Vasas* besättning – samt några båtsmäns äventyr i Stockholm. *Marinarkeologisk Tidskrift* 25.4: 7–9.

Sea Articles 1570: *Then Swenske Siörätt och Skieps Articklar Ao 1570 av Hertigh Carl. No 1*.

Skenbäck, Urban, 1983: *Sjöfolk och knektar på Wasa*. Vasa Studies 11. Stockholm.

Soop, Hans, 2001: *Silver, brons, mässing, tenn: Bruksföremål från örlogsskeppet Vasa*. Stockholm.

Söderlind, Ulrica, 2006: *Skrovmål: Kost-hållning och matlagning i den svenska flottan från 1500-tal till 1700-tal*. Stockholm.

Villner, Katarina, 1986: *Blod, kryddor och sot: Läkekonst för 350 år sedan*. Vasa Studies 14. Stockholm.

SINKING

Borgenstam, Curt, and Sandström, Anders, 1984: *Why did Wasa capsize?* Vasa Studies 13. Stockholm.

During, Ebba, 1994: *De dog på Vasa: Skelett-fynden och vad de berätta*. Vasa Studies 16. Stockholm.

Hafström, Georg, 1958: Örlogsskeppet Wasas undergång 1628. *Tidskrift i Sjöväsendet* 121.11: 740–770.

Kessler, Eric H., Bierly III, Paul E. and Gopalakrishnan, Shanthi, 2001: *Vasa* Syndrome: Insights from a 17th-century new-product disaster. *Academy of Management Executive*, 15.3: 80–91.

Squires, Arthur M., 1986: *The Tender Ship: Governmental Management of Technological Change*. Boston.

CONSEQUENCES

Chapman, Fredrik H. af, 1806: *Försök till en Theoretisk Afhandling att gifwa åt Linie-skepp Deras Rätta Storlek och Form likaledes för Fregatter och Bevärade mindre Fartyg*. Karlskrona.

Glete, Jan, 2002: Gustav II Adolfs Äpplet. *Marinarkeologisk tidskrift* 25.4: 16–21.

Glete, Jan, 2010: *Swedish naval administration 1521–1721: Resource flows and organizational capabilities*. Leiden.

Marinstaben, 1937: *Sveriges sjökrig 1611–1632. Särtryck ur Generalstabens Sveriges krig 1611–1632*. Stockholm.

Probst, Niels M., 1996: *Christian 4.s flåde*. Marinehistoriske skrifter. Copenhagen.

Zetterstén, Axel, 1891: *Svenska flottans historia, åren 1522–1634*. Stockholm.

LOST BUT NOT FORGOTTEN

Ahnlund, Nils, 1920a: Vikstensfyndets gåta löst. *Svenska Dagbladet* 20 July.

Ahnlund, Nils, 1920b: Ett storhetstids-minne i Stockholms hamn, *Svenska Dagbladet* 29 August.

Ahnlund, Nils, 1920c: note on the *Riksnyckelns* ordnance, in Underrättelser, *Historisk Tidskrift 1920*.

Cassel, Bo, 1957: *Dykning och dykare*. Stockholm.

Cassel, Bo, 1977a: Dykarkonstens utveckling. *Sjöhistorisk Årsbok 1975–1976*: 9–36.

Cassel, Bo, 1977b: Utvecklingen går vidare. *Sjöhistorisk Årsbok 1975–1976*: 37–56.

Fälting, Per Edvin, 1977: Tungdykare under 1900-talet. *Sjöhistorisk Årsbok 1975–1976*: 65–96.

Grisell, B., and Ahlberg, S., 1973: *Riksnyckeln 1628*. Forum Navale Skriftserie nr 28.

Hafström, Georg, 1958: Äldre tiders bärgningsarbeten vid vraket av skeppet Wasa. *Tidskrift i Sjöväsendet* 121.11: 771–844.

Hafström, Georg, 1961: A.L. Fahnehjelm och skeppet Vasa. *Tidskrift i Sjöväsendet* 124.6: 451–481.

Halley, Edmund, 1717, The art of living under water. *Philosophical Transactions* 29 (for the years 1714–1716): no. 349, pp. 492–499. London.

Hamilton, Edward, 1957: En marinarkeologisk undersökning utförd av Statens sjöhistoriska museum. *Sjöhistorisk Årsbok 1955–1956*: 163–183.

Jonsson, Per (ed.), 2003: *Skärgårdens bottnar. En sammanställning av sedimentundersökningar gjorda 1992–1999 i skärgårdsområden längs svenska ostkusten*. Naturvårdsverket, Rapport 5212. Stockholm.

Kjellin, U., 1997: Vatten och avlopp i kretslopp. In *S:t Eriks årsbok 1997*, pp. 7–32. Uppsala.

Kuylenstierna, O., 1921: Vikstenskanonerna nu utställda. *Hvar 8 Dag. Illustrerat Magasin* 23.8 (20 November).

Negri, Francesco, 1700: *Viaggio settentrionale in otto lettere*. Padua.

Stackell, Lennart, 1920a: Bronskanoner från vår stormaktstid upptagna från havsbotten. *Hvar 8 Dag. Illustrerat Magasin* 21.4 (1 August).

Stackell, Lennart, 1920b: Regalskeppet *Riksnyckelns* kanoner och vrakrester. *Hvar 8 Dag. Illustrerat Magasin* 22.4 (24 October).

Stackell, Lennart, 1921: Bärgningen av "Riksäpplet". *Hvar 8 Dag. Illustrerat Magasin* 23.3 (16 October).

Triewald, Mårten, 1734: *Konsten at lefwa under Watn ...* Stockholm. Published in English as *The Art of Living Under Water*, together with Triewald 1741, by the Historical Diving Society, London, 2004.

Triewald, Mårten, 1741: *Plägning til konsten at lefwa under Watn ...* Stockholm. Published in English with Triewald 1734 as *The Art of Living Under Water*, av Historical Diving Society, London, 2004.

SALVAGE AND EXCAVATION

Cassel, Bo, 1960: Marinen och Wasa. *Tidskrift i Sjöväsendet* 124.12:1112–1124.

Claus, Gillis, 1986: *Wasas historia 1956–64: upptäckt, bärgning, utgrävning*. Vasa Studies 15. Stockholm.

Franzén, Anders, 1957: Örlogsskeppet *Vasa*. Marinarkeologi i Stockholms hamn. *Tidskrift i Sjöväsendet* 120.

Franzén, Anders, 1961: Vasa. *Regalskeppet i ord och bild*. Stockholm.

Franzén, Anders, 1963. The Warship "*Vasa*". *The American-Scandinavian Review* 51.1 (March): 13–26.

Fälting, Per Edvin, 1961: *Med Vasa på Strömmens botten*. Falköping.

Fälting, Per Edvin, 1990: *I vått och torrt*. Stockholm.

Hallvards, Bengt, 1964: Pusslet Wasa. *Tidskrift i Sjöväsendet* 127.9: 590–597.

Håfors, Birgitta, 2001: *Conservation of the Swedish warship* Vasa *from 1628*. Stockholm.

Håfors, Birgitta, 2010: *Conservation of the the Wood of the Swedish warship* Vasa *of A.D. 1628: Evaluation of Polyethylene Glycol Conservation Programmes*. University of Gothenburg, Gothenburg Studies in Conservation 26, 2010.

Lundström, Per, 1963: *Utgrävningen av Wasa*. Vasa Studies 2. Stockholm.

VASA TODAY

Almkvist, Gunnar, 2008: *The Chemistry of the* Vasa: *Iron, acids and degradation*. Swedish University of Agricultural Sciences, Uppsala.

Fors, Yvonne, 2008: *Sulfur-Related Conservation Concerns for Marine Archaeological Wood: The origin, speciation and distribution of accumulated sulphur and some remedies for* Vasa. Stockholm University.

Sandström, Magnus, Fors, Yvonne and Persson, Ingmar, 2003: *The* Vasa's *New Battle: Sulphur, Acid and Iron*. Vasa Studies 19. Stockholm.

Picture sources

Many new photographs were taken for this book by Anneli Karlsson of the National Maritime Museums of Sweden. Other illustrations were produced by Medströms Bokförlag in cooperation with the *Vasa* Museum. Older images come largely from the publisher's archive or the National Maritime Museums. Otherwise, the following sources provided images:

Armémuseum, Stockholm 50 (top)
Bergting, Peter, Stockholm 10–11, 100–101
Deen, Mathijs, Amsterdam 36
Evans, Mark, Stockholm 136
Falu Gruva, Falun 48
Gezelius, Malin, Stockholm 19, 28, 30, 66, 147
Grönlund, Catarina, Gammelstad 67
Haninge kommun (photographer: unknown) 169
Kalmar läns museum, Kalmar 53
Kalmar Nyckel Foundation, USA (photo: Andrew Hanna) 82
Kempe, Claus, Mönsterås 32
Krigsarkivet, Stockholm 128, 129
Livrustkammaren, Stockholm 26 (bottom)
Magnusson, Roine, Roineimages 18
Magnusson, Thomas, Stockholm 203
National Maritime Museum, Greenwich 54
Riksarkivet, Stockholm, 38, 40, 137, 138–139 (photo: Kurt Eriksson),
Rosenborgs slott, Copenhagen 149
Schück, Hjalmar, Stockholm 37, 44, 60, 141
Seeberg, Annette, Hornslett 86
Seijbold, Olle 195
Sjöfartsverket (05-03013) 127
Stockholms stadsmuseum 50 (bottom)
Tiefenbacher, Richard 193

Index

acid 166, 199, 201
"Adam" 113, 131
Agamemnon 154
Ahnlund, Nils 168
Älvsborg fortress 25
Älvsborg Ransom 25
Älvsnabben 65, 102, 121, 123
Amsterdam 22, 36
anchorsmith 35
Äpplet (of 1619–25) 93, 145
Äpplet (of 1629–59) 65, 145, 146, 147, 148, 153, 155
armament 52, 55
armament plans 55
arming cloths 112
arrende system 141, 144
Atlas 181
Augustus (Roman emperor) 69

Bälinge (church) 80
ballast 133, 134, 135, 136, 138, 190
Baltijsk (Pillau) 29
Banér, Johan 144
barber-surgeon 104, 119
Batchvarov, Krum 200
beakhead 73, 78, 112, 193
"Beata" 113, 131, 132
beating 105
Bechstadius, Carl 167
Beckholmen 125, 126, 127
Belos 179, 180
Bengtsson, Sven 187
Bertil, Prince 176
Bertilsson, Per (boatswain) 103, 138
Besche, Willem de 161
Birger Jarl 74
Biržai (Birsen) 57
Biskopsholmen 125
Blasieholmen 35, 44, 124
boarding axe 63
boatswain 103
boltrope 83

bonnet 92
bowsprit 89
brace, bracing 92
Breitenfeld, battle of (1631) 143
Brömsebro, Peace of (1645) 153
bronze 51
Broström company 176
bulkhead 108, 109, 110
Bulmer, Ian 128, 160, 161
buoyancy 133

Calvin, Jean 19
canister shot 63
cannon 49–50, 51, 52, 53, 55–56, 57, 58
cannon founding, manufacturing 50
capstan 96, 97, 107, 121
Carlbom, Lennart 15
carpenter's store 108, 187
carpenters 33, 34, 37, 43, 44, 45, 81, 104
carvers 68, 79, 80, 81
carving 67, 77, 79, 81

Cassel, Bo 180
catheads 72
ceiling 12, 45
Central Maritime Museum, Gdańsk 60
centre of effort (of sail) 86, 88
centre of gravity 133
Cerberus/Kerberos 74
"Cesar" 113, 131, 132
chain shot 61
chainwales 87, 111, 112
chaplain 104
Chapman, F.H. af 155
Charles I (England) 128
cherubs/cherubim 71, 104
Christian II (Denmark) 24
Christian IV (Denmark) 30, 31, 128, 144, 148, 149, 153
Christina 148, 161
Clason, Edward 175, 176, 179, 182, 184
Clausink, Hans 73, 80
Clerck, Hans 83, 90, 91, 140
Clerck, Rickard (admiral) 83, 90, 91
clews 92, 122
climate system 198
coach 106, 111
colours 76–79
conscription system 94
conservation 192, 199–201
contracts 36–37, 38
cook 104
Copenhagen (København) 22, 30, 64, 154
Copper Company 50
coring device 169, 173
corrosion 166, 188
crew 62, 63, 95, 97, 99–119
culverin 51, 53, 56

Dahlbäck, Marianne 195, 196
Dalarö 169
Danzig (see Gdańsk)
David (Biblical king) 67, 72

De Geer, Louis 50, 140
demi-cannon (see 24-pounder) 49, 53, 54
diet 117
Dirschau (Tczew), battle of (1627) 26
disease 118
diving bell 157, 161, 162, 164
diving dress 167
Djurgården (Stockholm) 125, 195
DNA analysis 113
dolphins (on guns) 52, 57
Domesnäs (Kolgas rags) 39
Draken 150
Dutch war of independence (80 Years War) 41

elevation (of guns) 61
emblems 69
"Eric" 131
erosion bacteria 165

Fahnehjelm, Anton Ludwig 167
Fälting, Per Edvin 15, 16, 171, 173, 174, 175, 179, 184, 185
Falun copper mine 24, 49, 50
fascine/*fasces* (*vase*) 69
Fehmarn, battle of (1644) 143, 150–151, 152
figurehead 67, 69, 72, 73, 78
"Filip" 114, 131, 132, 189
Fleming, Henrik (vice admiral) 65, 117
Fleming, Klas Larsson (vice admiral) 28, 33, 39, 138, 152
Folkunga dynasty 71
food 106, 107
footropes 96
Forbes, Alexander 161
fore course, foresail 85, 86, 122
foremast 85
fore topsail 87
form stability 133
frames 45, 200

Frankfurt an der Oder, conquest of (1631) 55
Franzén, Anders 15, 169, 172–174, 176, 179, 184, 185
Friberg, Stig 15
Frigg 179, 180, 183, 185

Gabbard, battle of (1655) 54
Galärvarvets naval cemetery 141
Galion de Guise 39
galley (hearth) 106, 108, 118
galley (ship type) 29
Gamla Svärdet 65
Gdańsk (Danzig) 23, 33, 64
German Church (Stockholm) 67, 80
Gessus, Medardus 49, 50, 55, 56, 67
Giärdt, carver 80
Gideon (Biblical leader) 68, 75
Gierdsson, Petter (lieutenant) 83, 84, 102, 136
gilding 16, 76, 77
Göteborg (Gothenburg) 28, 29, 147, 159, 162
Göteborg 151
great cabin 100–101, 106, 110, 111
Great Copper Mountain (*Stora Kopparberget*) 14, 49
Great Power Period 7, 16, 23, 26
griffon 71, 75
Groot, Arendt de (see Hybertsson)
grotesques 70, 75
gun carriage 52, 56, 189
gun crew 104
Gun Foundry (Stockholm) 50
gundeck 31, 39, 134
gundeck, lower 8, 10–11, 55, 58, 107, 109, 130, 162, 189
gundeck, upper 106, 109, 186
gunmetal 50
gunners 63, 104
gunports 7, 51, 55, 59, 126, 133, 134, 154

INDEX 209

gunpowder 51, 52
"Gustafsson" (see "Helge")
"Gustav" 131
Gustav I Vasa (Gustav Eriksson) 24, 27, 28, 47
Gustav II Adolf 25–26, 27, 28, 31, 39, 40, 47, 50, 53, 55, 56, 57, 64, 65, 67, 69, 71, 72, 75, 111, 143, 144, 147, 148
Gustaf VI Adolf 17
Gustavus 37
Gyllenhielm, Karl Karlsson (admiral of the realm) 28, 38, 136

Habsburg 20
Hamilton, Edward 179
Hammer, Johan 144
Hanseatic League 24
Hansson, Söfring (captain) 83, 102, 119, 121, 127, 135, 141, 148, 161
headrope 93
heavy divers 13
Hedberg, Axel 176, 180, 184
Hedenäset (church) 67, 80
"Helge" ("Gustafsson") 14, 130, 131, 132
helmsman 72, 89, 110
hemp 42, 83, 93
Henrik Hybertsson (see Hybertsson)
Henrik Jacobsson (see Jacobsson)
Henry Grace à Dieu 39
Hercules (Herakles) 72, 74
herms 80
Hidemark, Ove 196
Hinnersson, Mikael 193
Hocker, Emma 201
hold 105, 107, 108, 130, 133, 190
Holy Roman Empire 19
Horn, Gustaf 144
Horn, Paridon von 93, 94
hull 34, 40, 43, 44, 45, 46, 133, 134
Hybertsson brothers (Henrik and Arendt) 33, 34, 36, 41, 44, 51, 90, 93
Hybertsson, Arendt (de Groot) 33, 36

Hybertsson, Henrik 33, 36, 40, 42, 83, 160
hygiene 119

inquest 136, 137
Isbrandtsson, Johan 42, 139
isotope analysis 114
"Ivar" 114, 115, 131, 132

Jakobsson, Henrik "Hein" 42, 44, 46, 138, 140, 141, 146, 147
Jansson, Ragnar 15
"Johan" 115, 129, 130, 190
Johan III 175
Johan Casimir, Duke 43
joiners 44, 81
Jönsson, Erik (vice admiral) 55, 65, 104, 111, 126, 136, 140
Jonsson, Hans (captain) 84, 102, 115, 126, 128, 129, 130
Juhani, diver 157–159

Kalmar Nyckel 82
Kalmar Union 24
Kalmar war (1611–13) 146
Karl, Duke (later King Karl IX) 27, 71, 75
Karl IX 27, 75
Karl X Gustav 120
Karlskrona ropewalk 91
kartog 53
Kastellholmen 125
King David 95
Kirkholm, battle of (1605) 21
knights/knightheads 71, 90
Kolberger Heide, battle of (1644) 148, 149, 155
kollegium system 28
Königsberg (Kaliningrad) 41
Krabbe, Erik 128, 144
Krommenie (Holland) 94
Kronan (fortification) 28

Kronan (of 1632–71) 65, 128, 147, 152, 153, 155
Kronan (of 1668–76) 53, 169

ladder 10, 108
lampblack 77
leading seaman 103
leading seaman's mate 104
Liberton, Jörgen 164
lieutenant 102, 104, 111
Lindormen 150
line tactics 154
Livonia 23, 27, 41
longboat 126, 179
lower mast 84, 85
Lübeck 22, 27, 164
"Ludvig" 129, 131
Lützen, battle of (1632) 26
Lundström, Per 17, 186, 188, 190
Lundvall, Bo 176
Luther, Martin 19

Magnus, Ole 90
main course/mainsail 85, 86, 92, 93
main topsail 86, 121, 124
mainmast 84, 191, 193
malachite 79
Månsson, Göran 195, 196
Margareta Nilsdotter 33, 36
mariners 102
maritime archaeology 201
Marstrand, Wilhelm 148
Mårten Redtmer (see Redtmer)
master 102, 104
master gunner (*konstapel*) 63, 104
master's mate 103
matchlock 62
Matsson, Jöran (master) 83, 84, 102, 103, 122, 138, 140, 161
Maule, Jacob 162
Maurice of Nassau, Prince of Orange 20

Medardus Gessus (see Gessus)
medicine 116
Meerman 150
merchant ship 96
mess 99, 104
messenger 97
mizzen 86, 87, 88, 121, 122, 124
mizzen bonnet 85
mizzenmast 88, 197
mizzen topsail 87, 88
Monier, Anton 36
Mund, Pros (admiral) 151
Municipal Court building, Stockholm 80
musket 62, 63

National Archive 139
national coat of arms 19, 71, 76
National Heritage Board 16, 176, 186
National Maritime Museum (Sweden) 16, 169, 174, 176, 179, 183
navy (Swedish) 16
navy yard (Stockholm) 7, 28, 29, 31, 33, 34–35, 36, 37, 43, 83, 124, 145
Negri, Francesco 161
Nelson, Horatio (admiral) 154
Neptune Towing and Salvage Company (Neptunbolaget) 16, 176, 179, 180, 181, 183, 196
Nilsdotter, Margareta 33, 36
nobles 111
non-commissioned officers 96, 103
Nordic Seven Years War (1563–70) 28
Nya Fortuna 150
Nyberg, Sven-Olof 15

Oden 179, 180, 183, 185
officers 99, 111
Oldenburg 150
Oldenburg dynasty (Denmark) 24
Oliwa, battle of (1627) 31, 60, 63, 143, 149
Olschanski brothers 168
Olschanksi, Simon 168

Öresund 149, 153
Öresund, battle of (1658) 62
Öresund toll 64, 153
orlop 85, 106, 108, 130, 187, 190
Övertorneå (church) 67
Oxdjupet 146, 147, 153
Oxenstierna, Axel 28, 40, 128, 148, 149
Oxenstierna, Gabriel 128

Palmstiernas Mekaniska Verkstad 173
Patientia 151
Peckell, Andreas 158, 159, 162, 164
Peenemünde 145
Petersz, Bonaventura 123
Petter, carver 80
Petter Gierdsson (see Gierdsson)
pikes 63
piling (for mooring) 121
Pillau (Baltijsk) 29
pilot 103
pinnace 29
Pipping, Olof 193, 203
Polish War (1599–1629) 25, 27, 64
polychrome 77
polyethylene glycol (PEG) 192, 195, 198, 201
Posse, Britta 43
powder magazine 108
provisions, provisioning 107, 108, 117
provost 104
punishment 105

quarter (crew division) 104
quarter galleries 58, 77, 110, 111, 129
quartering wind 92
quartermaster 104

Radziwiłł, Krzysztof 57
ramrod 62
rapiers 63
ratlines 122

Redtmer, Mårten 67, 68, 76, 79, 80, 81
Reformation 19, 20
regalskepp 147
Regina 151
rickets 117
riders 45, 46
Riga 21, 41, 42
Riga, conquest of (1621) 23, 27, 71
rigging 83, 84, 86–87, 90, 91, 93, 94–97, 193
Rijswijk (Holland) 36
Riksäpplet 168
Ripa, Cesare 68
robband 93
rope 83, 90, 91, 93
ropewalk 91, 93, 94
round shot 61
Royal Council 127, 160
"Rudolf" 130
running rigging 87, 96, 97
Ruyter, Michiel de (admiral) 153

sailcloth 94
salvage 15, 159, 171, 180–181
sandglass 105, 203
sawyers 34, 44
Scanian War (1675–79) 164
Scepter 147, 149, 152–155
scissor shot 61
sculptor 44, 68
sculpture 67–81
scupper 9
scurvy 117, 118
seamen 99, 103, 132
Septimius Severus (Roman emperor) 73
sheet 92, 122
Ship Company (Skeppskompaniet) 144
ship of the line 155
ship's boys 104
shipbuilding 43, 45
shipworm 146, 165

shipyard, Västervik 29, 93
shipyards 43
Sigismund (Zygmunt III) 26, 27, 30, 31, 67, 71
"Sigurd" 130
single-decker 39
sinking 121–161
Skeppsholmen (Blasieholmen) 35, 44, 124
Skokloster 152
Skytte, Johan 42
Sleipner 181, 183
Slussen (Stockholm) 123, 124
Smålands Lejon 150
Solen 60, 63
Spens, James 128
spike shot 61
spritsail 86, 88, 89, 112
spritsail topmast 89
spritsail topsail 86, 89, 112
square sail 84
Stackell, Lenny (commander) 168, 169, 173
Stadsholmen (Gamla stan, Stockholm) 124
steerage 70, 72, 89, 106, 110, 131, 132, 189
stern 74–75
stern gallery 110
sterncastle 112, 167, 193
steward 104
Stockholm 21, 123
Stockholm harbour 125
Stralsund, siege of (1628) 64, 65
streamers 91, 106, 111
Strömmen (Stockholm) 9, 125
sulphur 200
Svärdet 128
Svensson, Sam 179, 182
swivel guns 55, 110
swords 63

tacking 92
tankard 117

Tegelviken (Stockholm) 9, 123, 125
Thesson, Johan 80, 81
Thirty Years War (1618–48) 20, 25, 26, 144, 145, 149
three-decker 154, 148, 159, 161, 162–164
3-pounder 55, 56, 58, 111
Tiberius (Roman emperor) 69
Tigern 60, 149
tiller 88, 89, 109, 191
topgallantmast 84, 85
topgallantsail 85, 86, 87
topmast 85, 88
topsails 85, 86, 87, 96, 122
"Tore" 130
Torstensson, Lennart 144, 153
Torstensson's War (1643–45) 60
trade 24, 25
Trafalgar, battle of (1805) 154
train oil huts (Slussen, Stockholm) 121, 124
transom 71
Tre Kronor 39, 40, 67, 81
Tre Kronor palace 34, 124, 136
treenail 45
Trefoldigheden 148
Treileben, Albrecht von 159, 161, 162–164
Triewald, Mårten 162
tritons 73, 77, 78, 110
24-pounder 37, 49, 53, 55, 56, 59, 61, 71, 108, 134, 146, 155, 162
two-decker 154, 155

upper cabin 103, 106, 110
Uppsala cathedral 19
Usedom 144

Valdemar 73
Vasa Committee 174, 176
Vasa grave 141, 190
Vasa Museum 195–203
Vaxholm fortress 146

Viby estate (Sollentuna) 37
victims (died in sinking) 112–116, 130–131

Wallenstein, Albrecht von 31
warship 29
Wasa Board 176, 186
Wasa Shipyard 192, 195, 196, 197
watch (division of crew) 96, 103, 104
wearing ship 92
weather decks 111
weight distribution 133
weight stability 133
Welshuisen, Christian 93, 94
whipstaff 70, 72, 89, 90, 106, 110, 131
Wismar, conquest of (1632) 149
Wrangel, Karl Gustav (vice admiral) 150, 152, 153

Yk, Cornelis van 45
"Ylva" 130

Zetterström, Arne 177
Zetterström nozzle 177